HOW TO GROW A THRIVING
VEGETABLE GARDEN

✿ COUNTRYSIDE BOOKS

Publisher
Richard W. Morey

Assistant Publisher
Jeffrey A. Morey

Art Director
Jeffrey A. Morey

Production Manager
Gerald Pritzl

Administrative Services
Diana W. Morey

Computer Services
Cheryl Morey

Consultant
John T. Morey

Address all inquiries to:
Countryside Books
2280 US Hwy 19 North
Suite 223
Clearwater, FL 33575
813-796-7337

Copyright© 1984
ISBN 0-88453-0388

No portion of this book may
be reproduced without written
permission from the publisher.

Photography
Tom Eltzroth

Typography & Printing
Graphic Communications
Center, Inc.
Appleton, WI

Color Separations
Color-Tec, Inc.
Milwaukee, WI

A. B. Morse Co.
2280 US Hwy 19 North
Suite 223
Clearwater, FL 33575
813-796-7337

About the Cover
Photography concept by John T. Morey
Graphic Design by Jeffrey A. Morey

Acknowledgments

No author is infallible, especially when it comes to gardening information. Such information is not stored in data banks, partly because there are more exceptions than hard and fast rules.

Fortunately, the gardening field is blessed with regional experts and specialists in certain vegetables. They, too, are fallible, but their collective accuracy is a cut above that of writers who rely on personal experience and library research for the mass of their input.

Countryside Books and the authors acknowledge with thanks many people who checked copy for technical accuracy and some who contributed certain sections for each vegetable. Special thanks go to:

Thomas Eltzroth, Co-author, of San Louis Obispo, California, a Professor of Horticulture and an ardent home gardener. Tom collaborated closely with the author and publisher on the planting zone map and planting date chart, working from tried and true data from local areas all over the county. He also researched optimum plant spacing, supporting structures and storage. Most of the photographs in this book were shot by Tom in his own home garden.

Robert Barry, PhD, of University of Southern Louisiana. Dr. Barry was especially helpful on sweet potatoes, watermelon and okra.

George Brookbank of Tucson, Arizona, a keen home gardener. George contributed valuable information on gardening under desert conditions.

Milo Burnham, PhD, of Mississippi State University Extension Service. Dr. Burnham copy southern vegetables and contributed much of the material on southern peas.

Robert Johnston, Jr. of Johnny's Selected Seeds, Albion, Maine. Rob provided valuable assistance on vegetable culture for short season areas.

James Matheson of the Agway Research Station at Prospect, Pennsylvania. Jim was especially helpful on squash, melons, tomatoes and peppers for the Midwest.

T. L. Murdock, PhD, of the University of Arkansas at Fayetteville assisted with advice on watermelon.

Ray Rothenberger of Missouri State University at Columbia, Missouri. Ray reviewed a large group of vegetables including asparagus, cucumbers, endive, rhubarb, summer squash, swiss chard and tomatoes.

Glenn Rydl, PhD, at the Agricultural Department, Southwest Texas State University at San Marcos, Texas advised on crops grown in the Southwest.

Richard Schneider of the University of Wisconsin Extension Service. Dick reviewed cultural directions for upper Midwestern Climates and soils.

Joseph Steinke, a veteran horticulturist with the Rutgers South Jersey Research Station at Bridgeton. Joe called on years of experience in plant breeding and vegetable trials to review certain species in depth.

Paul Thomas of the Petoseed Co., Inc., research station Woodland, California. Paul verified copy on melons, tomatoes and peppers but, of course, can't be held responsible for changes made later in the copy. Plant breeders are perfectionists by nature and are understandably uncomfortable about the editing process.

My wife deserves special thanks for cheerfully tolerating the long hours of seclusion necessary to write a book that, although short in length, is long on accuracy and original material.

INTRODUCTION

Before pen is put to paper, or the word processor switched on, the publisher of a garden book must attempt to pinpoint its major audience so the author can write to it.

But, after reading an advance sample chapter of this book, Countryside Books could see that it had something for everyone who is interested in vegetable gardening. Even though it can be easily understood and applied by beginners, it does not have the paint-by-numbers simplicity that bores advanced gardeners. Conversely, advanced information is stripped of much of the gardening jargon that necessitates the learning of a new vocabulary.

The publisher invites you to skim a chapter, any chapter, and is confident you will find it enlightening, entertaining and above all, accurate.

ORIGINALITY

This book differs from other food gardening manuals in its starting point and its originality. Other books either grow out of one person's experience in gardening (often quite limited regionally) or are hatched by a committee of journalists working for a large corporation. Experiential books usually incorporate, without questioning, much existing information on "how-to" and "why to" and color it with the author's gardening travails and discoveries. Committee written books are usually carefully researched, journalistically clean and graphically exciting. But neither are original and both overlook or bypass the largest source of valuable information for home gardeners.

EXPERTISE FROM COMMERCIAL GROWERS

In the USA and Canada, commercial growers of vegetables have long been served by Extension Specialists and Research Scientists at the various state universities. This is a small circle of like minds but it is not a closed corporation. Information is open to all, but few outside the circle avail themselves of it. One drawback to adapting commercial grower techniques and research to home gardens is that very few people are sufficiently knowledgeable in both to glean the commercial fields for home garden ideas.

The author, for more than thirty years, has worked the worldwide interface between home and commercial food growing. He called on many specialists in the Cooperative Extension Service, Professors of Horticulture and experts in the garden seed industry for help. Spirited discussions were frequent and not all were happy with all the compromises.

Frequently, this blue ribbon panel found it necessary to question traditional techniques that seemed illogical. Eventually, this led to the discovery of an astounding number of errors, contradictions and omissions in respected reference books. This made them redouble their effort to make this book accurate. Yet, they would be the last to claim that information in it is infallible. Gardening is anything but black and white. The gray areas required many judgement calls. Some of these will be challenged.

AMERICA'S MOST AUTHENTIC GARDEN BOOK

Thus, by starting from a new base — betwixt commercial growers and home gardeners — and by taking the time to research every statement, the panel emerged with one of the most original garden books in many years.

Beginners will find especially useful the descriptions of plants, spacing and thinning directions, data on the mature size of plants and the number of plants needed per person. Such information should greatly simplify garden planning.

Advanced gardeners will see at once that this is truly a national book. While not addressing such local problems, for example, as how to grow vegetables in the alkaline soil of El Paso, it does incorporate much information gathered from the four corners of the country. Typical of its thoroughness is the seed planting and transplanting chart. It not only tells you when to plant in the various climate regions, but also when *not* to plant.

This book does not deal with the major problem facing most gardeners . . . insect pests and diseases. Actually, the information was compiled, but it proved so voluminous that it would have filled a separate book. Someday it may.

Countryside Books is confident that this book will speed your progress while helping you avoid gardening failures . . . and this will lead to your getting more satisfaction out of it. We hope our book will be your companion as you grow in gardening. If it becomes dog eared from use we will know that we achieved what we set out to do.

ABOUT THE AUTHORS

Jim Wilson - Author

Currently Jim Wilson and wife Jane own and operate Savory Farm, a commercial herb growing operation serving southern restaurants and gourmet food stores with fresh culinary herbs. Jim is also the host for the Southern Victory Garden Educational TV program seen in the Sunbelt and filmed at famous Callaway Gardens in Georgia.

From 1975 to 1982 he was Executive Secretary of All-America Selections and National Garden Bureau, Inc. The position entailed the promotion of home gardening and new award winning vegetable and flower varieties to the American public. During 1981 and 1982 he wrote and narrated 100 radio tape segments on home gardening under the "America's Gardener" tag. These were distributed to 200 stations in the USA and Canada and resulted in many radio and TV appearances where he encouraged people to garden and to use All-America selections in their gardens. He has been the keynote speaker or lecturer at a number of international horticultural meetings and workshops.

From 1967 to 1975 he was Advertising and Promotion Director and International Sales Representative for Sierra Chemical Company, Milpitas, CA, manufacturers of Osmocote Controlled Release Fertilizer. Wilson served Ferry-Morse Seed Co. from 1948 to 1967 after graduating from college in August, 1948.

Jim's education includes a BS in Agriculture, University of Missouri, 1948, Cum Laude. No advanced degrees, but many continuing courses in marketing, journalism, photography, psychology and art, over the years. His honors include Past National President, Garden Writers Association of America; Past National Vice President, Mens Gardens Clubs of America; Board of Directors, Marigold Society of America; Past Board Member, American Horticultural Society; International Award, Bedding Plants, Inc.; GB Gunlogson Medal, American Horticultural Society, for creative use of new technology in home garden design; Lecturer, the Williamsburg Gardening Symposium, 1981; Lecturer, the Callaway Gardens Symposium, 1981 & 1982.

Tom Eltzroth - Co-Author/Photographer

Tom Eltzroth grew up in southern Ohio where one of his first endeavors was helping his father in the family flower and vegetable garden. He studied horticulture at the Ohio State University where he received a Master of Science degree.

Tom moved to the central coast area of California in 1967, joining the horticulture faculty of California Polytechnic State University in San Luis Obispo. Along with his wife and daughter, Tom enjoys maintaining two large vegetable gardens each year - one is a year 'round garden at home, the other is a summer plot at a local community garden. Because of his close contacts with many of the major seed companies, Tom is able to observe first-hand the performance of most vegetables and flowers developed for home gardens .

In addition to his teaching at Cal Poly, Tom is a freelance garden writer and photographer. He regularly writes a column for *Flower & Garden* magazine and for two horticultural trade magazines. Additionally, his articles and photos appear in other horticultural publications. He has developed an extensive horticultural library which is the source of most of the photos used in this book.

TABLE OF CONTENTS

EASE OF GROWTH RATING

In the introduction to each vegetable, you will find a numerical "Ease Of Growth Rating," on a 1 to 10 scale. The easiest to grow are rated 1, the hardest to grow, 10. Most come somewhere in between.

These ratings apply in most, but not all, gardening situations. "Difficult" climates, for example, might affect ratings. In general, the following factors entered the rating for each vegetable:

Degree of gardening skill required to grow the vegetable well.

Speed of seed germination, particularly under adverse conditions.

Seedling vigor, and tolerance for less than perfect soil.

Rapidity of growth.

Tolerance for crowding.

Ability to endure heat and humidity.

"Holding power" — the ability to store on the vine vs. a requirement for prompt harvest.

SOIL PREPARATION

If you want a vegetable garden you can be proud of, start with thoroughly prepared soil. Most gardeners know this, and it is amazing how many will, year after year, settle for mediocre gardens rather than invest a little extra time and money in soil improvement.

Proper soil preparation will result in soil improvement, great at first, then gradual. Improved soils drain and warm up faster, soak up rather than shed water, and remain granular rather than crusting and baking hard. The improvement will also be reflected in ways you can't measure without sophisticated tests — in increased soil fertility levels and biological activity.

So, before you pick up a spade or crank up a tiller to prepare your garden soil, review this chapter to learn why good preparation pays off, as well as how to go about it.

THE IDEAL GARDEN SOIL

Some garden soils are nearly perfect but you won't, in a single season, be able to create one boasting:

Moderately fast drainage with good moisture retention.

An open, granular composition.

A high content of organic matter in fine particles.

A soil pH level adjusted to suit most vegetables.

A balanced level of major nutrients with deficiencies of micro-nutrients corrected.

Some gardeners are willing to make the necessary investment to build a good, if not ideal, garden soil in one season; others prefer to do it gradually.

PHYSICAL, BIOLOGICAL AND CHEMICAL IMPROVEMENT

The three sides to soil improvement are interdependent and are satisfied by careful soil preparation.

PHYSICAL IMPROVEMENT means improving the texture and structure of the soil and, through these, its drainage and aeration. Texture is the proportion of fine, medium and coarse particles in the soil; structure is their arrangement, either loose, dense and tight, or in between.

Physical improvement can be achieved by incorporating, through spading or tilling, soil amendments such as organic matter, sand, perlite, vermiculite, gypsum and lime. At the risk of belaboring the point, it must be stressed that no lime should be applied unless its need is indicated by a soil test.

CHEMICAL IMPROVEMENT comes through correcting deficiencies of major nutrient elements and micro-nutrients and by adjusting the soil pH to a level preferred by most vegetables, roughly between 5.8 and 7.0, with 6.0 to 6.5 the optimum range.

The major element phosphorus, and the compounds lime and gypsum are usually incorporated throughout the rootzone at the time soil is prepared. Phosphate and calcium tend to stay at the levels where you place them in the soil.

Gypsum contains calcium and sulfur but the main advantage in applying it to clay soils is the improvement of soil structure by a chemical loosening process.

BIOLOGICAL IMPROVEMENT means bettering the environment for beneficial soil organisms as small as fungi and bacteria and as large as earthworms. It comes about partly as a result of improved physical and chemical condition of the soil and, to a large degree, by the addition of organic matter.

While you can grow vegetables on soils devoid of organic matter, it isn't easy and it isn't particularly satisfying. It isn't good stewardship, a fundamental tenet of gardeners who want to improve their soil while reaping good harvests.

With so few places on earth uncontaminated by pollution and injudicious applications of chemicals, most gardeners take pride in and zealously guard their plots of healthy garden soil.

CLAY, SAND OR SILT, AND LOAM SOILS

Clay soils range from the red soils of Georgia to the black alkaline adobe of California and the grey or yellow subsoils of eastern suburbs, stripped of topsoil before development. Clays are stubborn, sticky and plastic when wet, and brick-hard when dry. They are cordially despised by gardeners who don't appreciate the redeeming feature that clays can be made into valuable, highly fertile soils, retentive of nutrient elements, capable of supplying moisture ·to crops long after sandy soils have dried out.

Clays are rarely all-clay but may contain sand, silt, gravel and stones. If only one third of the soil particles by weight are clay, the soil will act like clay. Therefore, the job of physical improvement may be huge and is best done in installments.

Clay soils are usually clay all the way down to bedrock.Some, laid down by ancient lake or stream action, may be underlain by layers of gravel or sand or may have rock-like concretions in layers at various depth, which stop the movement of water and oxygen.

If your garden soil is clay, you have to decide whether to garden in it or on top of it. Gardening on top of clay soils is an easy but expensive route. In effect, you use the clay mostly as a reservoir for moisture and nutrients and overlay it with a bed of loamy soil. You start by spading or tilling the clay deeply and, at the same time, incorporating phosphate sources and lime if needed. On top of the tilled soil you build up beds of a mixture of sand, organic matter and not more than 25 percent of clay from the foundation soil.

How deep to make the beds depends on how much you can afford to spend for sand, manure, sawdust, pulverized pine bark, rice hulls or compost. These additions make a loose, well-drained seedbed where seeds and transplants can sprout or root easily with no crusting.

Gardening in clay soils can become very physical and frustrating. The soil should be worked only when it is dry enough not to pack into plastic lumps yet not so dry that it cracks into hard, sharp-edged chunks. Improving clay is usually a slow process unless you are willing to double-dig the soil to a depth of 18 to 24 inches, incorporating large amounts of organic matter and gypsum in the lower layer and sand and organic matter in the top 3 to 6 inches of soil.

The ultimate goal with clay soils is to have a porous, reasonably fast draining layer of highly amended soil to "spade depth", roughly 8 to 9 inches, underlain with clay that has been fortified with phosphate, lime (if needed), and enough organic matter and gypsum to let roots and water penetrate.

Large-scale gardeners improve clay soils by growing and turning under green manure crops at least yearly, or by composting and turning under leaves and grass. Roots open channels to lower layers and earthworms mix organic matter throughout the upper soil mass.

Some of the most productive garden soils are in the clay class and are noted for their nutrient and water holding power. Even when improved, clays will stick to your shoes and garden tools but it is a small price to pay for productivity.

SAND, SILT OR GRAVEL SOILS

These soils are dry and infertile and have a low capacity to store nutrients and water. Sand and gravel are loose except for fine sands which tend to pack solid. Silt tends to pack hard and to shed water. Adding clay to such soils is not recommended. The fine particles of clay slip through the sand or gravel and collect around obstructions. There, they form layers that impede drainage and aeration. Adding clay to silt creates a sticky, gritty "problem soil."

Organic matter can make sands, silts and gravels productive but must be regularly replenished, particularly on deep sands. As organic matter decomposes to fine particles of humus, these gravitate down to lower layers of sand and are lost.

An ideal situation for soil improvement is a layer of sandy soil underlain by clay or clay loam at a depth of 1 to 2 feet. In preparing such soils for a garden, avoid mixing the clay with the sandy topsoil except in small increments. Instead, shovel the topsoil aside and then turn over the clay, incorporating organic matter, gypsum and, if needed, lime and phosphate. Replace the sandy surface layer and mix in organic matter and lime as indicated by soil tests. Gypsum has no effect on the texture and structure of sandy soils.

Sandy soils warm up quickly and are preferred for early crops. Deep sands tend to be "droughthy" and require frequent applications of water during dry seasons. Nutrients are leached out by the water and have to be replaced frequently.

LOAM SOILS

Loams contain fractions of clay, sand, silt and, usually, organic matter. The loams range from clay loam which is hardly discernible from pure clay, to fine sandy loam which looks and acts much like sand. The loams are the easiest of the soils to prepare and keep in good physical and biological condition. When preparing loam soils underlain with clay, be careful not to dig too deep or you can drastically alter the physical characteristics of your topsoil.

If your soil is loamy, you are fortunate because you already have what other gardeners have to work for years to achieve — good "tilth", which shows as an easily worked soil. Other gardeners have to create a loam on top of their existing soil.

SPADING AND TILLING BARE SOIL (NOT TURF)

All gardens need to be dug yearly, except those operated after the late Ruth Stout's deep straw mulch philosophy. Digging breaks up compactions, restores aeration, incorporates such additives as lime, phosphate and organic matter and makes the gardener feel useful.

Digging your first garden may be a joy or an ordeal, depending on whether your soil is clay, and on your timing. If you try to attack hard, dry clay, even with a powerful tiller, you may be repulsed or, if you beat it into submission, dissatisfied with the result. If you dig clay or heavy clay loam that is wet enough to be plastic, you will wreck its structure for several reasons.

SPADING

The best implement for turning over the soil by hand is a strong, well made spade with a round point. (These are also called "shovels"; dealers are as confused as everyone else about the terminology.) Most gardeners select the long-handled model but some like the shorty with the "D" handle.

In spading, move laterally as you spade, and take small "bites" to ease the lifting burden. Push the spade deeply into the soil before levering back on the handle. Pick up the bite of soil with the blade and invert it in the spot where you removed it. Move one step to the side and take another small bite until you have worked a straight line across the garden. Each time you spade a line across the garden you invert another 6 to 9 inches of soil. Keep your spade blade clean with a wooden stick, and sharpened with a file or coarse emery stone. Soil builds up quickly on blades and makes digging much harder.

Now, you have inverted all the soil but you are not through. You need to spread organic matter, phosphate, wood ashes, sand and, if needed, lime over the spaded area. Then, go back over it and dig in the amendments to spade depth. You will find that the twice-over digging goes fast and results in deeper, more thorough digging than if you try to combine digging and incorporation of additives in one operation. Remove sticks, roots and clumps of grass. Whack stubborn clods with the back of the spade to pulverize them.

You are ready now to mark off rows or to make raised beds or ridged up rows. If your soil is sand or sandy loam, garden "on the flat." Little is gained by building up beds because these soils already drain well and warm up fast. If yours is a heavier soil, string a taut-line across the garden to mark off beds 3 to 4 feet wide. Walk the line to mark off a foot path then remove the line and shovel soil from the pathway to build up beds on either side of it.

TILLING

Most gardeners find that tillers are not economical for small plots of 300 to 500 square feet or less in area, unless they also have substantial beds of flowers, berries or orchard areas. If you have your heart set on a tiller, rent one first to see if it delivers what you expect. If you plan to use a tiller for cultivating paths between rows you will need to space rows wider apart than for hand cultivating, thus reducing the total productivity of your garden.

For larger gardens, tillers are a blessing. If your garden is clay or clay loam, pay extra and get the 8 h.p. model. The 5 h.p. and smaller models are adequate for sandy soils or clay soils that have been improved by the addition of organic matter over the years.

For the best results in tilling, wait for 4 or 5 days after a soaking rain. Test a handful of soil by squeezing it. If the soil remains in a ball that can't be crumbled easily, it is too wet to work. A moist but not wet soil is less likely to tear up the tines on a tiller if it contains rocks and debris. Moist soil will also work up into a desirable blend of small and medium-sized particles.

Dry clay will break into chips of hard soil, along with quite a bit of dust. With the first watering the fine clay particles will flow into a paste that will dry into a solid crust. Seedlings can't penetrate it and it will shed water. The effects of tilling dry clay soil will be evident in poor soil structure for several years.

Tillers do not dig as deeply as ads might imply. With tines measuring 10 inches tip to tip, a 5 inch tilled depth could be expected. Yet, this is sufficient for mixing layers of organic amendments and chemicals into improved garden soil. On new clay soil, till the soil in successive layers to a depth of 12 to 18 inches, shoveling each tilled layer aside. Mix the soil with amendments as you return it to the trench. Replace the subsoil first.

PREPARING GARDENS FROM SOIL IN TURFGRASS

Experience has shown that you are way ahead if you remove the turf before tilling or spading. Slice the turf into strips and pile it, roots up, in layers. Scatter manure or fertilizer on each layer and moisten the pile as it builds up. The turf will decompose in a few months to produce high-grade compost.

Removing sod greatly reduces problems with weeds, frees the soil of clumps of vegetation, and reduces the chance of nitrogen starvation on vegetables planted on soil containing the decaying vegetation. Above all, you won't have to fight the regrowth from clumps of sod and sprigs of persistent grass and weeds. Gardeners have found that scalping the soil clean of turfgrass down to a depth of 1½ inches will give them a virtually weed-free garden the first season.

Persistent perennial grasses such as bermuda, St. Augustine, quackgrass, centipede and tall fescue can be sprayed with a specific grass killing herbicide and allowed to die before tilling. On such deep rooted grasses, herbicides work much better than scalping the sod. Your County Agent/Farm Advisor can recommend a herbicide that will do its job thoroughly and quickly degrade into harmless residues. If you are intent on digging or tilling in grass, consider this treatment. Even if you shake out and remove every clump or sprig of roots from a turfgrass area,

millions of grass seeds and pieces of roots will remain to sprout. You will be fighting weeds and grass for years to come if you do not remove or kill the turf.

Be sure to work in a generous amount of organic matter to compensate for the removal of the sod. You will soon discover that the more you add, the easier it is to pull the few weeds that come up.

DRESSING THE SOIL

After tilling or spading the soil you will need to work it into what gardeners call "a good seedbed." This combines hoeing to break up clods and smoothing with an iron rake to make a level seedbed. The ultimate goal in dressing the soil is to produce a level, smooth seedbed with a texture like coffee grounds.

You will sometimes see experienced gardeners "gandy dancing" on beds of roughly spaded or tilled soils before they rake it level. Or, they will let a rain settle the roughly worked soil before smoothing. Either way, tramping the soil or settling it with water, will result in a bed that will stay level and that will sprout seeds fast. Spaded or tilled soils fluff up when worked. Unless it is settled before seeding, the excessively loose soil will interfere with the capillary movement of soil moisture that helps to sprout seeds.

Tramping might seem destructive to soil because it contradicts advice against soil compaction. But, experience has shown that, if you avoid tramping excessively moist soil, it will work to your benefit. A labor saving variation for large gardens is tramping down only the soil over seeded and covered rows, leaving the rest of the bed "as is."

CONTOURING ROWS OR BEDS

If your garden is on a slope, however slight, consider running your rows or beds across the slope to reduce erosion and to trap water until it can soak in.

On long rows don't trust yourself to "eyeball" contours. Stretch a taut line from end to end and hang a bubble-type line level in the center. This simple device will help you mark off footpaths or furrows that follow contour lines around slopes. A drop of 1 to 2 inches per 100 feet of row or bed will slow down runoff water and divert it off one end of the rows. In rainy areas you may have to slow down the water pouring out of the ends of furrows. Break up a bale of hay or straw and line the waterway with sheaves. They will serve as baffles.

The first rain may show water collecting in low sinks, in furrows or footpaths. After the puddles have dried, shovel soil into them from high spots in the furrows.

FEEDING PLANTS

Pity the beginning gardener trying to decide whether to feed his or her vegetable garden and, if so, how much, when, and how often. A typical garden center might offer boxes and bags of fertilizers from a half dozen manufacturers and, within those labeled for vegetables, several choices of analyses. And, to further complicate matters, two or three analyses of fertilizer might be needed to meet the requirements of the various vegetables.

Reading the chapter in this book on SOIL IMPROVEMENT would be a good place to start to clear the air. When you have well-drained soil, well fortified with organic matter, and you have adjusted the pH to the range preferred by vegetables, matters such as the analysis of fertilizer, application rate and frequency become less critical. When your soil has been improved to this happy state, you will find that a little fertilizer will go a long way.

A beginner would not go wrong if he or she followed the manufacturer's directions on the fertilizer package. But, these directions are, of necessity, very general. The manufacturer often recommends lean rates rather than to contribute to the chance of burning plants with excessive applications.

Most gardeners want to know the "whys" of fertilizing or feeding plants rather than just following directions and it is hoped that this chapter will supply enough answers without information overkill.

SOIL TESTS

Soil fertility is dynamic. Not one soil in ten will grow good vegetables through a full season without the addition of plant nutrients in some form. Heavy feeders such as sweet corn or potatoes may begin to show signs of nutrient deficiencies part way through the season. Regular soil tests, perhaps every two years, will give you benchmarks on how your soil improvement program is progressing as well as the recommended rates of fertilizers for the various vegetables. In most states, soil testing is provided at low cost by the agricultural college through local County Agricultural Agents, with offices usually in the County Seat. Agents are also known as "Farm Advisors" or "Cooperative Extension Agents". Their phone number can usually be found in the directory under the name of the university in your state where the College of Agriculture is headquartered.

Soil tests are not expensive and are necessary to good soil management. Some states now recommend organic fertilizers as well as manufactured products.

FEEDING THE SOIL

Some experts maintain that it is better to "feed the soil" and let it feed the plants. Whatever name the process is given, it should be done in concert with loosening and enlivening the soil with organic matter to improve drainage and the environment for beneficial soil organisms.

A well-drained, biologically active soil will make the best use of any fertilizers you apply. In a healthy soil, large and small organisms gradually convert fertilizers, soil particles and organic matter into nutrients that can be absorbed by plant roots. Fine particles of organic matter and soil attract and hold excess nutrients in a "soil bank" reserve.

FERTILIZERS AND SOIL AMENDMENTS

Fertilizers are concentrated sources of major plant nutrients. Soil amendments are sources of secondary nutrients, organic matter and bulk to improve the texture and structure of soil. Limestone, gypsum, peatmoss, sawdust, sand, Perlite and Vermiculite are amendments and do not carry analysis labels like fertilizer.

The fertilizers you buy in retail stores may contain one, two or three of the major elements: nitrogen, phosphorus and potassium. The three majors are always listed in the following order: NPK. The letters represent the plant nutrients nitrogen, phosphate and potash. In nature, phosphorus and potassium are unstable and cannot exist without combining with other elements to make compounds. They are always listed as "phosphate" and "potash" on USA fertilizer labels.

The percentage by weight of the major nutrient elements must be listed on fertilizer labels and is called the "analysis". Here are some examples:

33-0-0 Ammonium nitrate, 33% nitrogen.

0-46-0 Triple superphosphate, 46% phosphate.

0-0-60 Muriate of potash, 60% potash.

5-10-10 A typical "complete" fertilizer containing 5% nitrogen, 10% phosphate and 10% potash.

You may, in certain states, see minor elements such as iron, copper, boron, etc. added to fertilizers to correct common soil deficiencies. Generally, micronutrients are in adequate supply except on soils that are very sandy or alkaline.

A feature of fertilizers that puzzles beginners is why one can't buy brands with 100% "active ingredients". Why, for example, can't one purchase a 20-40-40 analysis? The answer is partly in the fact that phosphate and potash are chemical compounds containing oxygen molecules and/or salts or bases. These add weight and reduce the percentage of active ingredients. Practically speaking, the most concentrated complete analysis you can purchase easily is water soluble 20-20-20.

The slow and incomplete water solubility of most dry, granular fertilizers comes as a surprise to most new gardeners. These fertilizers are formulated to be worked into the soil where they gradually melt. Some have to be converted to available forms by soil organisms. If you prefer to "liquid feed" your plants with fertilizer dissolved in water, you will need to buy water soluble crystalline fertilizers or liquid concentrates. With the exception of nitrate, water soluble fertilizers usually cost significantly more than dry granular analyses.

WHICH ANALYSIS SHOULD YOU BUY FOR VEGETABLES?

For figuring nutrient needs, vegetables can be roughly divided into three classes:

Leafy vegetables: The salad greens and potherbs require about twice as much nitrogen as either phosphate or potash because of the large amount of green matter that must be produced in a relatively short time.

Fruiting vegetables: These are grown principally for their pods or fruits and usually require more phosphate and potash than nitrogen.

Root vegetables: These require all three major nutrients but especially potash. Some, such as turnips, like high levels of phosphate in the soil.

Two analyses of manufactured fertilizers should suffice for your vegetables:

10-5-5 or 10-10-10 for leafy vegetables. You may prefer to use higher analysis fertilizers with the same 2:1:1 NPK ratio. You should apply less, of course, because they are more powerful and apt to burn plants.

5-10-10 for fruiting and root vegetables. A higher analysis fertilizer with the same 1:2:2 NPK ratio would be 10-20-20.

Some knowledgeable gardeners with large plots prefer to work with straight sources of nitrogen, phosphate and potash and lay in a bag each of ammonium nitrate or ammonium sulphate, superphosphate and either muriate of potash or potassium nitrate.

You will find that phosphate needs are the most difficult to predict. Soils in the humid South are almost always deficient in phosphate and, to grow good vegetables, will need pre-plant as well as supplementary applications during the season. Soils in the arid West are often low in *available* phosphate because, in alkaline soils, phosphate combines with the calcium and sodium in the soil and is not available to plants . . . they can't absorb these insoluble salts. The pH of such soils may have to be lowered to make soil reserves of phosphate available to plants.

ORGANIC FERTILIZERS

Some gardeners prefer to use fertilizers made from animal or vegetable by-products or unprocessed ores. These are good but, per unit of plant nutrients, expensive. Organic sources of nitrogen abound. All release available nitrogen slowly in cool or cold soils but at a more rapid rate in warm soils due to the stepped up activity of soil organisms. Organic sources of phosphate are few and release very slowly; they are of questionable value on alkaline soils unless they are highly modified with organic matter. Organic potash sources are rare and are usually of low analysis.

Typical organic sources of nitrogen and, secondarily, phosphate and potash, are blood meal, tankage, fish emulsion and some of the stronger animal wastes such as chicken or sheep manure.

Some organic sources of phosphate are rock phosphate and bone meal. Except on acid soils, these release so slowly that they are of little value to plants the first year.

Greensand marl and cocoa bean mulch are used by organic gardeners as sources of potash.

Organic sources of phosphate and potash are not water soluble and work best if incorporated into the soil during preparation.

Organic gardeners sometimes unknowingly purchase publicized kinds of by-product fertilizers, not realizing that they may be fortified with chemical sources of nutrients. Manufacturers have trouble selling certain non-fortified organic fertilizers because the analyses are so low — 5-2-1, for example, or 2-0-0. Therefore, some 'juice up' the analysis with concentrated mineral fertilizers to produce a saleable product. This is legal but hard to detect. Packages and labels are sometimes so small that sharp vision is needed to read the sources of plant nutrients listed on the label, especially in the dim light of a garden center.

Promising research is underway to develop plants that, like members of the legume family, will extract nitrogen from the air. Similarly, it is conceivable that, through gene splicing, organisms could be developed that could ingest nutrient ores to produce efficient phosphate and potash sources acceptable to organic gardeners.

MANUFACTURED OR CHEMICAL FERTILIZERS

Most of the fertilizers sold in North America are manufactured by concentrating ores or by extracting nitrogen from the air. Processing fertilizers requires large amounts of energy but, with their relatively high analyses, less energy per unit of active ingredient is expended in shipping from factory to market.

Most manufactured fertilizers release nutrients quickly and can burn plants if applied at rates higher than recommended. Extended-release fertilizers are excellent for vegetables but, per unit of plant food, expensive. They are especially effective on sandy soils and the artificial soils used in container growing.

For home garden purposes, go by the analysis, not by the plant name on the box or bag. Manufacturers are always pushing for more shelf space and package many specialty fertilizers. Beans or lettuce won't know the difference if you feed them 5-10-10 from a bag labeled "Flower Garden Food."

Many gardeners are convinced that the highest yields of good quality vegetables can be obtained by combining the good stewardship of organic gardening with low to moderate applications of manufactured fertilizers. This approach is popular with the Cooperative Extension Service.

FERTILIZER APPLICATION RATES AND FREQUENCY OF FEEDING

Please read these notes before referring to the following fertilizer chart.

This chart is intended to supply only supplemental information as an aid to adjusting manufacturers' recommended rates to your soil, climate and crops.

Recommended rates in books and bulletins are usually keyed to the percent of nitrogen in mixed chemical fertilizers or animal by-products. For the sake of simplicity, we have chosen 10-10-10 as a representative mixed chemical fertilizer and cotton-seed meal, 6-2-1 analysis, as a typical organic fertilizer. These two analyses would be good for supplemental feeding of leafy vegetables or, in moderation, for fruiting and root vegetables.

The recommended rates and frequencies are based on supplementation of phosphate and potash sources incorporated in the soil as pre-plant fertilizers. For best garden performance, use preplant fertilizers each time you turn over the soil, plus supplemental fertilizers at the recommended rates and frequencies.

If your soil is heavy, poorly drained clay or hard-packed silt and you fail to loosen it with organic matter and build up raised beds, use the low rates listed. In dense or soggy soil the nitrogen fraction in fertilizers remains longer in the ammoniacal form and can burn plant roots.

PREPLANT APPLICATIONS OF FERTILIZER

Skilled gardeners usually mix fertilizers into the soil along with organic matter and, if needed, lime or gypsum. Digging or tilling in calcium, phosphate and potash sources places them in the rootzone of plants. Calcium and phosphorus remain at levels where you place them, gravitating down very slowly. Potassium is somewhat more mobile, especially in sandy soils. Only nitrogen in the nitrate form will move freely through the soil.

Organic gardeners often prefer one of two approaches to preplant fertilization as part of their soil maintenance program:

Scattering manure, rock phosphate, greensand marl or wood ashes over a green manure crop and turning it under for summer crops.

Applying manure or compost plus rock phosphate and either greensand marl or wood ashes to bare soil and turning it over in the spring for early crops.

Green manure crops, if the winter is mild and the spring early, may develop a large mass of green matter that is slow to decompose. Bare ground is usually ready earlier for spring crops.

Gardeners who prefer commercial fertilizers often topdress them, along with manure, over green manure crops in the fall. The organic matter helps to reduce the loss of nutrients to erosion and leaching. For new garden plots prepared in the spring, they scatter preplant fertilizers along with organic matter and spade or till it into the rootzone.

Mixed complete fertilizers such as 10-10-10 are usually preferred for poor soils rather than a combination of phosphate and potash sources alone. Spring application of nitrogen fertilizers is more efficient because nitrogen tends to leach away with winter rains and snow melt when applied in the fall.

FEEDING VEGETABLES IN CONTAINERS

The loose, fast-draining artificial soil mixtures such as commercial potting soil are excellent for growing vegetables but they contain little or no nutrients and some, designed for azalea growing, are highly acid. Formulators usually don't tell you whether the soil mixtures contain lime.

As a general rule, you should fortify potting soils with ¼ lb. ground limestone per cu. ft., preferably dolomitic, to provide sufficient calcium and magnesium to last the season. You can also mix in 1 to 2 oz. of triple superphosphate per cu. ft., or twice this amount of rock phosphate. Unless you fortify potting soils with calcium and phosphate, certain species may suffer from blossom-end rot and/or dwarfing.

Fortified potting soils can be kept productive with supplemental fertilizers containing all three major nutrients. Liquid fertilizers are especially popular for feeding container plants and the long-feeding controlled release fertilizers are gaining favor. Manure teas can also be used. To help you in converting, 1 cu. ft. of soil is equivalent to 7 gallons.

FREQUENCY AND RATES OF APPLICATION

For Supplemental Feeding With 10-10-10 Or Cottonseed Meal 6-2-1
Begin Supplementary Feeding 30-45 Days After Planting

Vegetable Crop	Soil Type							
Rates are expressed as pounds of fertilizer per 100 sq. ft. "1x" means feed once, "2x" feed twice.	Fertile, improved soil, moderately drained to well drained		Poor, unimproved soil, moderately drained to well drained		Dense, heavy or hard soil, poorly drained		Sandy or gravelly soil, well drained to fast draining (1)	
	10-10-10	6-2-1	10-10-10	6-2-1	10-10-10	6-2-1	10-10-10	6-2-1
Asparagus, Established 1x to 2x	1-2	2-4	2-4	4-6	1-2	2-4	2-4	4-6
Beans, Green Snap, Bush 1x	0-½	½-1	½-1	1-2	½	½-1	½-1	1-2
Beans, Green Pole or Runner 2x	0-½	½-1	½-1	1-2	½	½-1	½-1	1-2
Beans, Lima, Bush 1x	0-½	½-1	½-1	1-2	½	½-1	½-1	1-2
Beans, Lima, Pole 2x	0-½	½-1	½-1	1-2	½	½-1	½-1	1-2
Beets 1x	½-1	1-2	1-2	2-4	½-1	1-2	½-1	2-3
Broccoli 1x	½-1	1-2	1-2	2-4	½-1	1-2	1-2	2-3
Brussels Sprouts 2x	½-1	1-2	1-2	2-4	½-1	1-2	1-2	2-3
Cabbage 1x	½-1	1-2	1-2	2-4	½-1	1-2	1-2	2-3
Carrots 2x	0-½	½-1	½-1	1-2	0-½	½-1	½-1	1-2
Cauliflower 1x	1-2	2-4	2-4	4-6	1-2	2-4	2-4	4-6
Celery 1x to 2x (if needed)	1-2	2-4	2-4	4-6	1-2	2-4	2-4	4-6
Chinese Cabbage 1x	½-1	1-2	1-2	2-4	½-1	1-2	1-2	2-3
Collards 1x to 2x (if needed)	½-1	1-2	1-2	2-4	½-1	1-2	1-2	2-3
Corn, Sweet, Early 1x	½-1	1-2	1-2	2-4	½-1	1-2	1-2	2-3
Corn, Sweet, Late 2x	½-1	1-2	1-2	2-4	½-1	1-2	1-2	2-3
Cucumbers 1x to 2x (if needed)	½-1	1-2	1-2	2-4	½-1	1-2	1-2	2-3
Eggplant 1x to 2x (if needed)	1-2	2-4	2-4	4-6	1-2	2-4	2-4	4-6
Endive 1x	½-1	1-2	1-2	2-4	½-1	1-2	1-2	2-3
Kale 1x	1-2	2-4	2-4	4-6	1-2	2-4	2-4	4-6
Kohlrabi 1x	½-1	1-2	1-2	2-4	½-1	1-2	1-2	2-3
Leeks 1x	1-2	2-4	2-4	4-6	1-2	2-4	2-4	4-6
Lettuce, Leaf 1x	0-½	½-1	½-1	1-2	½	½-1	½-1	1-2
Lettuce, Head & Romaine 2x	0-½	½-1	½-1	1-2	½	½-1	½-1	1-2
Melons 1x	1-2	2-4	2-4	4-6	1-2	2-4	2-4	4-6
Mustard Greens 1x	½-1	1-2	1-2	2-4	½-1	1-2	1-2	2-3
Okra 1x to 2x (if needed)	½-1	1-2	1-2	2-4	½-1	1-2	1-2	2-3
Onions, for Scallions, also Shallots 1x	0-½	½-1	½-1	1-2	½	½-1	½-1	1-2
Onions, Bulbs, also Garlic 1x	½-1	1-2	1-2	2-4	½-1	1-2	1-2	2-3
Parsley 1x to 2x (if color fades)	½-1	1-2	1-2	2-4	½-1	1-2	1-2	2-3
Peas, English or Green 1x	0-½	½-1	½-1	1-2	½	½-1	½-1	1-2
Peas, Southern 1x	½-1	1-2	1-2	2-4	½-1	1-2	1-2	2-3
Peppers 1x, 2x if long season	½-1	1-2	1-2	2-4	½-1	1-2	1-2	2-3
Potatoes, Irish 1x	1-2	2-4	2-4	4-6	1-2	2-4	2-4	4-6
Potatoes, Sweet 1x	1-2	2-4	2-4	4-6	1-2	2-4	2-4	4-6

NOTES: (1) On fast draining soils feeding every 2-4 weeks may be required, depending on organic matter content and leaching of nutrients by rain or irrigation.

Vegetable Crop

Soil Type

Rates are expressed as pounds of fertilizer per 100 sq. ft. "1x" means feed once, "2x" feed twice.	Fertile, improved soil, moderately drained to well drained		Poor, unimproved soil, moderately drained to well drained		Dense, heavy or hard soil, poorly drained		Sandy or gravelly soil, well drained to fast draining (1)	
	10-10-10	6-2-1	10-10-10	6-2-1	10-10-10	6-2-1	10-10-10	6-2-1
Pumpkin 2x	½-1	1-2	1-2	2-4	½-1	1-2	1-2	2-3
Radish 1x	0-½	½-1	½-1	1-2	½	½-1	½-1	1-2
Rhubarb Established 1x	1-2	2-4	2-4	4-6	1-2	2-4	2-4	4-6
Rutabaga 1x	1-2	2-4	2-4	4-6	1-2	2-4	2-4	4-6
Spinach 1x	1-2	2-4	2-4	4-6	1-2	2-4	2-4	4-6
Squash, Summer 1x	½-1	1-2	1-2	2-4	½-1	1-2	1-2	2-3
Squash, Winter 2x	½-1	1-2	1-2	2-4	½-1	1-2	1-2	2-3
Swiss Chard 2x	½-1	1-2	1-2	2-4	½-1	1-2	1-2	2-3
Tomatoes, Dwarf or Compact 1x	½-1	1-2	1-2	2-4	½-1	1-2	1-2	2-3
Tomatoes, Intermediate 2x	½-1	1-2	1-2	2-4	½-1	1-2	1-2	2-3
Turnips 1x	1-2	2-4	2-4	4-6	1-2	2-4	2-4	4-6
Watermelon 1x	1-2	2-4	2-4	4-6	1-2	2-4	2-4	4-6

NOTES: (1) On fast draining soils, feeding every 2-4 weeks may be required, depending on organic matter content and leaching of nutrients by rain or irrigation.

Typical Preplant Fertilizers And Application Rates Designed For Use With Supplemental Fertilizers

Organic Fertilizers	Fertile, improved soils, moderately drained to Well drained	Poor, unimproved soil, moderately to well drained	Dense, heavy or hard soil, poorly drained	Sandy, gravelly soil, fast draining
Bonemeal 3-6-0	4	10	5	5
Rock phosphate 0-15-0	3	5	4	4
Wood ashes 0-1-7	2	5	3	3
Tankage 7-10-2	2	4	3	2
Manufactured Fertilizers				
Triple Superphosphate	2	3	2	2
Muriate of Potash	1	2	1	1
Basic Slag	4	6	5	4
10-10-10	2	4	2	2

Rates are expressed as pounds of fertilizer per 100 sq. ft.

UNDERSTANDING PLANTING DATES
AND THE ZONE MAP

Planting dates in each zone cover a time span from as short as just a few days to as long as several months. Within a set of springtime dates, the first date shows the earliest safe planting time without taking elaborate precautions against frost and cold. The last date indicates the time beyond which planting usually gives unsatisfactory results. Dates for other seasons should be viewed similarly, except the last date is usually closer to frost and cold weather, so the earlier date is often preferable for planting.

Some crops in certain zones show two planting time spans; indicating a crop can be started successfully during two seasons. Both seasons may be equally suitable; but for some, one planting season may be preferable. For example, collards and kale planted during the latter dates reach harvest during the cool weather of autumn or early winter, producing crops far tastier than those timed for summer harvest.

CONSIDER GROWING SEASON

When deciding on a planting time, consider whether the crop grows better during a cool season or a warm season. If it is a cool season crop, such as spinach, plant nearer the first date in spring and nearer the last date in summer or fall. Warm season crops, such as snap beans, should be planted nearer the last spring date and nearer the first summer or fall date.

Plantings made during late summer and fall usually mature more slowly than plantings made earlier because of the cooler weather and shortening daylengths of autumn. If possible, when fall planting, select quick maturing varieties; realizing that days to maturity as indicated on seed packs and in catalogs are only relative, and are based on growth rate during the summer months.

Length of the growing season is a major factor determining what crops and varieties grow successfully in an area. Remember that planting of most cold tolerant crops can begin before the last spring frost and harvest can extend beyond the first fall frost. The gardening season can be extended even further with the use of coldframes, cloche coverings, and other protective devices.

MEAN DATES USED

The zone map is based on the mean date of the last spring frost and the first fall frost. Within a zone, frost dates can vary by as much as 15 to 20 days both in the spring and in the fall. Consequently, length of the growing season (number of days from spring frost to fall frost) varies up to as much as forty days within a zone. The growing season in some northern and high elevation areas can be shorter than 100 days, or as long as 365 days in Texas and California southern coastal regions and in southern Florida.

Generally, when gardening near the southern reaches of a zone, it is safe to plant on or close to the first spring planting date and the last fall planting date. Near the northern limit of a zone, spring planting should be delayed until the last spring date indicated, and summer/fall planting should be close to the first summer/fall date. If in doubt, delay spring planting. Except when a crop requires a long growing season, a week or ten day delay in spring planting usually causes no problems.

Frost dates within a particular zone are not solely dependent on its latitude since elevation, air movement and environmental abnormalities also exert an influence. A garden located near a large body of water is often in a microclimate warmer than the zone map indicates. In such locations, spring planting can often be advanced by as much as two weeks, and fall planting dates can usually be extended. Elevation within a zone has a drastic effect on planting times. As a rule, a 1000 foot increase in elevation means the last spring frost will come 7 to 10 days later, and the first fall frost date will come 7 to 10 days sooner. This factor is reflected in the zone map; but not precisely enough to show subtle variations within some zones, particularly those from the Rocky mountains west.

STICK TO RECOMMENDED DATES

The best source of reliable frost date information is the local county extension office; but no one can predict precisely when frosts will occur. The mean frost dates used in preparing the zone map are averages, indicating there is a 50% chance frost will come earlier and a 50% chance it will come later.

Zone maps and planting dates are flexible; but beginning gardeners should stick by the recommended planting dates for the first few seasons, keeping accurate records of frost dates, planting and harvest dates for individual varieties, and making notes of climatic peculiarities in the garden. Eventually, you will get a fix on the climate and frost patterns in your area and will learn the special features of your own garden site.

COUNTRYSIDE PLANTING ZONE MAP

20

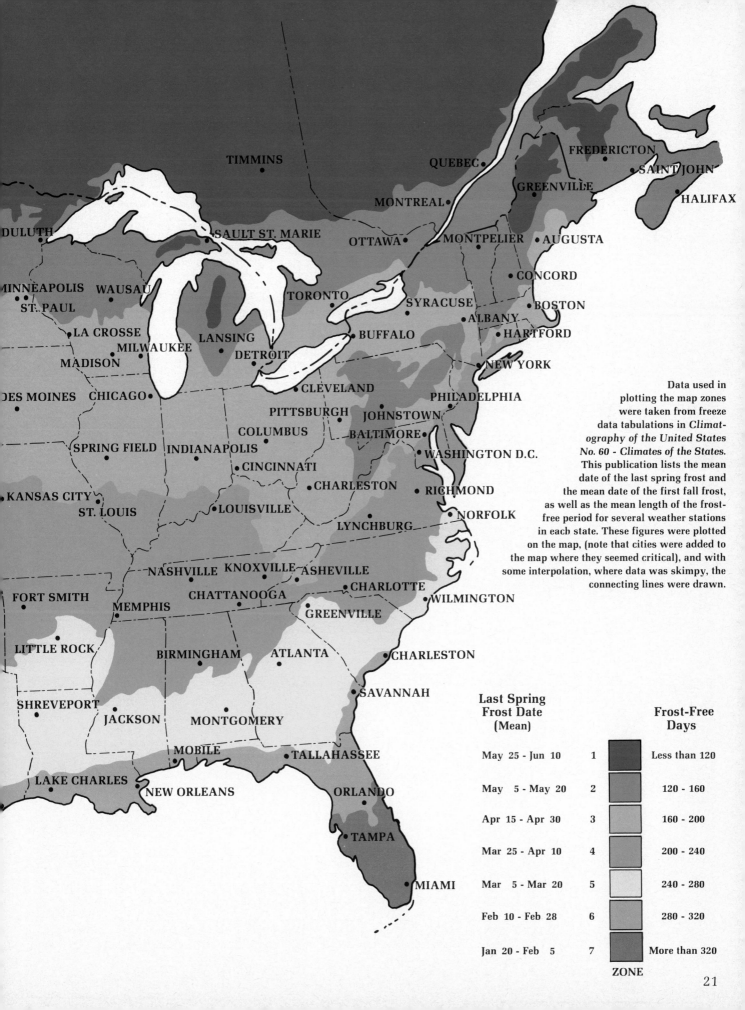

TIMMINS

QUEBEC

FREDERICTON

SAINT JOHN

GREENVILLE

MONTREAL

HALIFAX

DULUTH

SAULT ST. MARIE

OTTAWA

MONTPELIER • AUGUSTA

MINNEAPOLIS   WAUSAU

CONCORD

ST. PAUL

TORONTO

SYRACUSE

BOSTON

LA CROSSE

ALBANY

MILWAUKEE   LANSING

BUFFALO

HARTFORD

MADISON

DETROIT

NEW YORK

DES MOINES   CHICAGO

CLEVELAND

PHILADELPHIA

PITTSBURGH

JOHNSTOWN

COLUMBUS

BALTIMORE

SPRING FIELD   INDIANAPOLIS

WASHINGTON D.C.

CINCINNATI

KANSAS CITY

CHARLESTON

RICHMOND

ST. LOUIS

LOUISVILLE

NORFOLK

LYNCHBURG

NASHVILLE   KNOXVILLE   ASHEVILLE

FORT SMITH

CHATTANOOGA

CHARLOTTE

MEMPHIS

WILMINGTON

GREENVILLE

LITTLE ROCK

CHARLESTON

BIRMINGHAM   ATLANTA

SHREVEPORT

SAVANNAH

JACKSON   MONTGOMERY

MOBILE

TALLAHASSEE

LAKE CHARLES

NEW ORLEANS

ORLANDO

TAMPA

MIAMI

Data used in
plotting the map zones
were taken from freeze
data tabulations in *Climat-
ography of the United States
No. 60 - Climates of the States.*
This publication lists the mean
date of the last spring frost and
the mean date of the first fall frost,
as well as the mean length of the frost-
free period for several weather stations
in each state. These figures were plotted
on the map, (note that cities were added to
the map where they seemed critical), and with
some interpolation, where data was skimpy, the
connecting lines were drawn.

Last Spring Frost Date (Mean)		Frost-Free Days
May 25 - Jun 10	1	Less than 120
May 5 - May 20	2	120 - 160
Apr 15 - Apr 30	3	160 - 200
Mar 25 - Apr 10	4	200 - 240
Mar 5 - Mar 20	5	240 - 280
Feb 10 - Feb 28	6	280 - 320
Jan 20 - Feb 5	7	More than 320

ZONE

WHEN TO PLANT

Vegetable		Zone 1	Zone 2	Zone 3	Zone 4	Zone 5	Zone 6	Zone 7
Asparagus - using dormant crowns or seedlings	TR.	May 1-Jun 1	Apr 10-May 15	Mar 15-May 15	Feb 1-Mar 15	Jan 15-Mar 10	Nov 1-Mar 1	Nov 1-Feb 1
Bean - Dry types	D.S.	Jun 15-Jun 20	May 25-Jun 25	May 5-Jul 15	Apr 15-Jul 1	Apr 1-Jun 1	Mar 15-May 15	Feb 25-May 1
Beans, Green Bush types	D.S.	Jun 15-Jun 25	May 25-Jul 5	May 5-Jul 25	Apr 15-Aug 1	Apr 1-May 15 / Aug 1-Aug 15	Mar 15-May 1 / Aug 15-Sep 15	Feb 25-Apr 15 / Aug 25-Sep 25
Beans, Green Pole types	D.S.		May 25-Jun 10	May 5-Jun 15	Apr 15-Jul 1	Apr 1-Apr 20 / Aug 1-Aug 10	Mar 15-Apr 15 / Aug 1-Aug 10	Feb 25-Mar 15 / Aug 1-Sep 1
Beans, Lima Bush types	D.S.		Jun 1-Jun 15	May 10-Jun 20	Apr 20-Jul 1	Apr 5-Jun 15	Mar 20-Jun 1 / Aug 5-Aug 20	Mar 1-May 15 / Aug 15-Sep 5
	TR.		Jun 5-Jun 20	May 15-Jun 25	Apr 25-Jul 5	Apr 10-Jun 20	Mar 25-Jun 5	Mar 5-May 20
Beans, Lima Pole types	D.S.		Jun 1-Jun 10	May 10-Jun 10	Apr 20-Jun 15	Apr 5-Jun 5	Mar 20-May 15	Mar 1-May 5
	TR.		Jun 5-Jun 15	May 15-Jun 15	Apr 25-Jun 20	Apr 10-Jun 10	Mar 25-May 20	Mar 5-May 10
Bean - Shell types (For green shell use)	D.S.	Jun 15-Jun 20	May 25-Jun 25	May 5-Jul 15	Apr 15-Jul 1	Apr 1-Jun 1 / Aug 1-Aug 15	Mar 15-May 15 / Aug 15-Sep 1	Feb 25-May 1 / Aug 25-Sep 10
Beets	D.S.	May 10-Jul 1	Apr 20-Jul 20	Apr 1-Jul 25	Mar 10-Apr 20 / Aug 1-Sep 5	Feb 20-Apr 15 / Aug 20-Sep 20	Feb 1-Apr 1 / Sep 1-Oct 5	Sep 15-Mar 15
	TR.	May 20-Jul 1	May 1-Jul 20					
Broccoli	D.S.			Jul 1-Jul 25	Jul 10-Aug 5	Jul 20-Aug 20	Aug 1-Aug 20	Sep 1-Feb 1
	TR.	May 15-Jul 10	May 1-Jul 20	Apr 10-May 5	Mar 25-Apr 15	Mar 1-Mar 25	Feb 5-Mar 5	Sep 15-Feb 10
Brussels Sprouts	TR.	May 15-Jun 1	May 15-Jun 15	May 15-Jun 25	Mar 5-Apr 10 / Jul 10-Aug 10	Feb 15-Mar 25 / Jul 25-Aug 25	Jan 20-Mar 1 / Aug 10-Sep 10	Aug 20-Feb 5
Cabbage - Early crop	TR.	May 5-May 31	Apr 15-May 10	Mar 25-Apr 20	Mar 5-Apr 10	Feb 15-Mar 15	Jan 20-Feb 15	Jan 1-Jan 31
Cabbage - Second and Late Crops	D.S.		Apr 25-May 10	May 15-Jun 15	Jul 1-Aug 1	Jul 25-Aug 15	Aug 1-Sep 1	Aug 20-Nov 15
	TR.	May 10-May 31	May 10-Jun 10	Jun 15-Jul 15	Jul 10-Aug 10	Aug 10-Sep 1	Aug 15-Sep 15	Sep 1-Dec 31
Cabbage, Chinese	D.S.	May 10-Jun 25	Apr 25-May 20 / July 1-July 15	Apr 5-May 1 / Jul 10-Aug 1	Mar 20-Apr 10 / Jul 20-Aug 15	Feb 25-Mar 20 / Aug 1-Sep 1	Feb 1-Mar 1 / Aug 10-Oct 1	Jan 10-Feb 5 / Aug 20-Nov 1
	TR.	May 15-Jun 30	May 1-May 25	Apr 10-May 5	Mar 25-Apr 15	Mar 1-Mar 25	Feb 5-Mar 5	Jan 15-Feb 10
Carrots	D.S.	May 10-Jun 10	Apr 20-Jul 10	Apr 1-Jul 20	Mar 10-Apr 20 / Aug 1-Sep 1	Feb 20-Apr 15 / Aug 20-Sep 15	Feb 1-Apr 1 / Sep 1-Oct 1	Sep 15-Mar 1
Cauliflower	TR.	May 15-Jun 30	May 1-Jul 10	Apr 10-May 5 / Jul 1-Jul 20	Mar 25-Apr 15 / Jul 10-Aug 1	Mar 1-Mar 25 / Jul 20-Aug 15	Feb 5-Mar 5 / Aug 10-Sep 1	Jan 25-Feb 10 / Sep 15-Nov 20
Celery	TR.	Jun 10-Jun 25	May 20-Jun 10	Apr 30-May 25	Apr 10-Apr 30	Mar 20-Apr 1 / Jul 1-Jul 30	Feb 25-Mar 15 / Jul 25-Aug 15	Aug 15-Feb 15
Collards	D.S.	Apr 25-May 20	Apr 5-Apr 25	Mar 15-Apr 5 / Jul 10-Aug 10	Feb 25-Apr 1 / Jul 25-Sep 1	Feb 5-Mar 15 / Aug 1-Sep 15	Jan 10-Feb 20 / Aug 10-Oct 10	Sep 1-Jan 25
	TR.	May 5-May 30	Apr 15-May 5	Mar 25-Apr 15 / Jul 20-Aug 20	Mar 5-Apr 10 / Aug 1-Sep 10	Feb 15-Mar 25 / Aug 5-Sep 25	Jan 20-Mar 1 / Aug 15-Oct 15	Sep 10-Feb 5
Corn	D.S.	Jun 1-Jun 15	May 10-Jun 25	Apr 20-Jul 5	Apr 1-Jul 10	Mar 15-May 10 / Jul 10-Aug 10	Mar 1-Apr 30 / Jul 20-Aug 20	Feb 15-Apr 30 / Jul 20-Aug 30
	TR.	Jun 10-Jun 20	May 20-Jul 5					
Cucumber	D.S.	Jun 15-Jun 25	Jun 1-Jul 1	May 10-Jul 15	Apr 20-Aug 1	Apr 5-Jul 1 / Aug 15-Sep 1	Mar 20-Apr 20 / Aug 15-Sep 5	Mar 1-Apr 1 / Sep 1-Sep 25
	TR.	Jun 20-Jul 1	Jun 5-Jul 5					
Eggplant	TR.		Jun 10-Jun 20	May 20-Jun 20	May 1-Jun 15	Apr 10-Jun 1	Mar 25-Apr 20 / Jul 15-Aug 1	Mar 5-Apr 1 / Aug 1-Sep 5
Endive	D.S.	May 10-Jun 1	Apr 15-May 15	Mar 25-Apr 5 / Jul 10-Jul 25	Mar 10-Apr 1 / Aug 15-Sep 15	Aug 15-Sep 15	Jan 1-Feb 15 / Sep 1-Oct 10	Jan 1-Feb 1 / Sep 1-Oct 15
	TR.	May 25-Jun 10	May 1-May 20					
Kale	D.S.	Apr 25-May 20	Apr 5-Apr 25 / May 30-Jun 20	Mar 15-Apr 5 / Jul 15-Aug 15	Feb 25-Apr 1 / Aug 1-Sep 5	Feb 5-Mar 15 / Aug 5-Sep 20	Jan 10-Feb 20 / Sep 20-Oct 20	Sep 1-Feb 1
	TR.	May 5-May 30	Apr 15-May 5 / Jun 5-Jun 30	Mar 25-Apr 15 / Jul 25-Aug 25			Sep 1-Nov 1	Sep 10-Dec 31
Kohlrabi	D.S.	May 10-Jun 15	Apr 25-Jul 1	Apr 5-Jul 25	Mar 20-May 1 / Aug 1-Sep 1	Feb 25-May 1 / Aug 15-Sep 15	Feb 1-Apr 15 / Sep 1-Sep 20	Sep 25-Apr 15
	TR.	May 15-Jun 20	May 1-Jul 5	Apr 10-Aug 1	Mar 25-May 5	Mar 1-May 1	Feb 5-Apr 20	Jan 5-Apr 20
Leeks	D.S.	May 10-Jun 1	Apr 15-May 15	Mar 25-May 1	Mar 5-Apr 10	Feb 15-Mar 15 / Jul 15-Jul 25	Jan 25-Mar 1 / Aug 1-Aug 20	Jan 1-Feb 15 / Sep 5-Nov 5
	TR.	May 25-Jun 10	May 1-May 25	Apr 10-May 10	Mar 20-Apr 20	Mar 1-Mar 25 / Jul 25-Aug 5	Feb 10-Mar 10 / Aug 10-Sep 1	Jan 15-Feb 25 / Sep 15-Nov 15

D.S. - Direct Seed, TR. - Transplant

Vegetable		Zone 1	Zone 2	Zone 3	Zone 4	Zone 5	Zone 6	Zone 7
Lettuce - Looseleaf, Butterhead & Cos types	D.S.	May 10-Jun 30	Apr 15-May 20 Jul 25-Aug 5	Mar 25-Apr 30 Aug 1-Aug 20	Mar 5-Apr 20 Aug 15-Sep 10	Feb 15-Mar 31 Sep 1-Sep 30	Jan 25-Mar 31 Sep 10-Oct 15	Sep 10-Mar 31
	TR.	May 20-Jul 10	Apr 25-Jun 1	Apr 5-May 10	Mar 15-May 1	Feb 25-Apr 10	Feb 5-Apr 10	Jan 10-Apr 10
Lettuce - Crisphead types	D.S.		Jun 25-Jul 10	Jul 15-Aug 10	Aug 10-Sep 1	Aug 20-Sep 20	Sep 1-Oct 5	Sep 1-Dec 20
	TR.	May 25-Jun 30	Apr 30-May 20 Jul 5-Jul 20	Apr 10-Apr 30 Jul 25-Aug 20	Mar 20-Apr 20 Aug 20-Sep 1	Mar 1-Mar 31 Sep 1-Oct 1	Feb 10-Mar 1 Sep 10-Oct 15	Jan 20-Mar 31 Sep 10-Dec 31
Melons	D.S.		Jun 1-Jun 20	May 10-Jun 5	Apr 20-May 10	Apr 5-May 1	Mar 20-Apr 20	Mar 1-Apr 15
	TR.	Jun 15-Jun 20	Jun 5-Jun 25	May 15-Jun 10				
Mustard Greens	D.S.	Apr 20-Jul 10	Apr 1-Jul 25	Mar 15-Apr 30 Aug 1-Aug 20	Feb 10-Apr 15 Aug 10-Sep 5	Jan 15-Apr 1 Aug 20-Oct 5	Jan 1-Mar 15 Sep 1-Nov 10	Sep 15-Feb 15
Okra	D.S.				May 1-Jun 10	Apr 15-Jun 15	Apr 1-Jul 1 Aug 15-Sep 1	Mar 10-Jun 1 Aug 15-Sep 5
	TR.	Jun 15-Jun 25	Jun 10-Jun 25	May 30-Jun 15				
Onions	D.S.		Apr 20-May 15	Apr 1-Apr 25	Mar 10-Apr 15 Sep 1-Oct 10	Feb 20-Mar 20 Sep 10-Nov 5	Feb 1-Mar 1 Sep 15-Nov 15	Sep 15-Jan 15
	TR.	May 10-Jun 5	Apr 25-May 20	Apr 5-May 1	Mar 15-Apr 20 Sep 5-Oct 15	Feb 25-Mar 25 Sep 15-Nov 10	Feb 5-Mar 5 Sep 20-Nov 20	Sep 20-Jan 20
Parsley	D.S.	May 10-May 25	Apr 20-May 15	Apr 1-May 1	Mar 10-Apr 1	Feb 20-Mar 25 Aug 15-Sep 15	Feb 1-Mar 15 Sep 1-Oct 1	Sep 1-Mar 1
	TR.	May 25-Jun 10						
Peas	D.S.	May 1-Jun 25	Apr 15-May 15 Jul 1-Jul 15	Mar 15-May 1 Jul 10-Jul 25	Feb 10-Mar 20 Jul 20-Aug 5	Jan 20-Mar 15 Aug 25-Sep 5	Jan 10-Mar 10 Sep 10-Oct 10	Sep 15-Feb 25
Peas, Southern	D.S.			May 10-Jul 1	Apr 20-Jul 15	Apr 5-Aug 10	Mar 20-Aug 25	Mar 1-Sep 15
Peppers	D.S.						Jun 10-Jul 1	Jun 20-Jul 20
	TR.	Jun 15-Jun 20	May 25-Jun 20	May 5-Jun 20	Apr 15-Jun 15	Apr 1-May 20	Mar 15-Apr 20 Jul 20-Aug 15	Feb 25-Apr 1 Aug 1-Sep 5
Potato (Irish) - using transplants grown from true seeds	TR.	Jun 5-Jun 30	May 15-Jun 15	Apr 25-May 20	Apr 5-May 5 Aug 10-Aug 25	Mar 20-Apr 20 Aug 20-Sep 1	Mar 5-Apr 5 Sep 1-Sep 10	Feb 15-Mar 15 Sep 10-Dec 31
Potato (Irish) - using tuber 'seed' pieces	D.S.	May 10-May 20	Apr 20-Jun 1	Apr 1-May 5	Mar 10-Apr 1 Aug 1-Aug 15	Feb 20-Mar 15 Aug 10-Aug 20	Jan 25-Feb 20 Aug 20-Sep 1	Sep 1-Feb 15
Pumpkin	D.S.	Jun 10-Jun 20	Jun 1-Jun 20	May 10-Jun 15	Apr 20-Jul 1	Apr 5-Jul 1	Mar 20-Jul 1	Mar 1-Jul 1
	TR.	Jun 15-Jun 25	Jun 5-Jun 25					
Radish - Salad types	D.S.	May 5-Jul 25	Apr 15-Aug 15	Mar 25-Sep 1	Mar 1-Apr 25 Sep 1-Sep 30	Feb 10-Apr 10 Sep 1-Oct 25	Sep 15-Apr 1	Oct 1-Mar 15
Radish - Winter types	D.S.	May 25-Jun 25	Jun 15-Jul 10	Aug 1-Aug 15	Aug 15-Sep 1	Sep 1-Sep 20	Sep 15-Oct 10	Oct 1-Oct 20
Rhubarb - using crowns or divisions	TR.	May 5-Jun 1	Apr 20-May 15	Mar 25-May 1	Mar 5-Apr 10	Feb 15-Apr 1	Oct 15-Mar 1	Dec 1-Feb 5
Rutabaga	D.S.	Jun 1-Jun 20	May 15-Jul 5	Jul 1-Jul 25	Jul 25-Aug 25	Aug 10-Sep 5	Aug 15-Sep 10	Sep 1-Sep 25
Spinach	D.S.	May 5-Jun 5	Apr 10-May 10 Jul 15-Aug 20	Mar 15-Apr 20 Aug 10-Sep 15	Feb 10-Mar 15 Sep 1-Sep 30	Jan 15-Mar 10 Sep 5-Oct 20	Jan 1-Feb 15 Sep 15-Nov 5	Sep 15-Feb 10
Summer Squash	D.S.	Jun 10-Jun 25	Jun 1-Jul 1	May 10-Jul 10	Apr 20-Jul 20	Apr 5-May 20 Aug 1-Aug 15	Mar 20-Apr 25 Aug 10-Aug 25	Mar 1-Apr 10 Aug 10-Sep 5
	TR.	Jun 15-Jun 30	Jun 5-Jul 5	May 20-Jul 20	May 1-Jul 20	Apr 15-Jun 1	Apr 1-May 5	Mar 10-Apr 20
Sweet Potato - using rooted "slips"	TR.		Jun 1-Jun 5	May 15-Jun 15	May 1-Jun 20	Apr 15-Jun 20	Mar 15-Jul 1	Mar 1-Jul 1
Swiss Chard	D.S.	May 10-Jun 5	Apr 20-Jul 1	Apr 1-Jun 15 Jul 15-Jul 25	Mar 10-Jun 15 Jul 25-Aug 15	Feb 20-Jun 1 Aug 1-Sep 10	Aug 15-Apr 1	Sep 1-Apr 1
Tomatoes	D.S.			Apr 20-May 20	Apr 1-May 15	Mar 15-Apr 20 Jun 15-Jun 25	Feb 20-Mar 20 Jun 20-Jul 5	Feb 1-Mar 1 Jul 1-Jul 20
	TR.	Jun 15-Jun 20	May 25-Jun 20	May 5-Jun 20	Apr 15-Jun 15	Apr 1-May 20 Jul 15-Jul 25	Mar 15-Apr 20 Jul 20-Aug 5	Feb 25-Apr 1 Aug 1-Aug 20
Turnip	D.S.	May 5-Jul 1	Apr 15-May 15 Jul 10-Jul 25	Mar 25-May 1 Aug 1-Aug 15	Mar 1-Apr 1 Aug 15-Sep 15	Feb 10-Mar 25 Aug 20-Sep 30	Jan 1-Mar 25 Sep 1-Oct 15	Oct 1-Mar 15
Watermelon	D.S.		Jun 1-Jun 10	May 10-Jun 15	Apr 20-Jun 30	Apr 5-May 15	Mar 20-May 1	Mar 1-Apr 20
	TR.	Jun 1-Jun 5	Jun 5-Jun 15	May 15-Jun 20	Apr 25-Jul 5	Apr 10-May 20	Mar 25-May 5	Mar 5-Apr 25
Winter Squash	D.S.	Jun 10-Jun 15	Jun 1-Jun 10	May 10-Jun 15	Apr 20-Jul 1	Apr 5-Jul 1	Mar 20-Apr 10	Mar 1-Mar 15
	TR.	Jun 15-Jun 20	Jun 5-Jun 15	May 20-Jun 25	May 1-Jul 10	Apr 15-Jul 10	Apr 1-Apr 20	Mar 10-Mar 25

D.S. - Direct Seed, TR. - Transplant

ASPARAGUS

Asparagus

(No Common Names)

Asparagus officinalis

Liliaceae (Lily Family)

Cool Season

Perennial

Winter Hardy In Most Climates

Full Sun

**Ease Of Growth Rating, From Seeds -- 4;
From Crowns -- 3**

DESCRIPTION

Asparagus is grown for spears, which are the shoots that emerge from the soil in early spring. The spears grow from fleshy, spreading roots. Spears are cut or snapped off at ground level. On established plants, cutting can last for 6-8 weeks. As the cutting season progresses, the diameter and length of spears begins to decrease. That is the signal that the carbohydrates and nitrogenous reserves stored in the mat of the roots is nearly exhausted. It is time to let some of the spears grow into plumes and, within 2 or 3 weeks, to discontinue cutting altogether until the following spring.

Within a few weeks after cutting has ended, asparagus changes to a mass of feathery foliage on thin, wiry stems. Its needle-like leaves convert energy from the sun to carry on life processes and to rebuild the food reserves in the roots. By the season's end the plumes may reach 5 feet in height. In zones 1, 2, and 3 they should be left on the plants through the winter and clipped off about when grass begins to turn green the following spring. Elsewhere, the plumes should be removed after the first heavy freeze and disposed of as they can shelter a number of harmful insects and rust spores.

Asparagus is adapted to all parts of North America except for most of Alaska, Northern Canada, and the virtually frost-free areas of zone 7. It is easy to grow except in zones 6 and 7, where mild winters do not permit the plants to go dormant. In arid areas of these warm zones asparagus can be grown by letting plants go dry and dormant for a month or two in midsummer. Spears from such plants may be smaller than those from winter-dormant plants.

The plants can be started from seeds or from "crowns" -- roots grown from seeds planted the previous year. Each method has its advantages. Plants grown from seeds have little chance of being infected with fusarium wilt, a soil-borne diease that frequently afflicts asparagus. Also, you can select any variety you wish. When you buy dormant crowns, the varietal selection is limited but the advantage of starting with crowns is that you can advance the time from planting to first harvest by at least a year. Most gardeners opt for starting from crowns.

Asparagus plants may be either male or female, the female recognizable by its berries. Either produces spears, but male plants yield somewhat more since they are spared the drain of berry production. Because a plant must be in its second or third year to show its sex, most home gardens will contain a mixture of male and female plants.

Asparagus is native to parts of Russia, the Mediterranean region, and the British Isles. It was used for food by the Romans and other ancient peoples and was highly regarded for medicinal purposes. The early colonists brought it to North America and it is now naturalized along fence rows, rail rights of way, and country roadsides in many areas.

SITE

Asparagus is a long lived plant and should be placed at the back of the garden to keep it out of the way of foot traffic or garden preparation. Place it in full sun if possible, but where it will not shade adjacent vegetables or flowers. The plants are billowy; they can cast shade and will compete aggressively with nearby plants for water and nutrients.

SOIL

Asparagus will survive on rather poor soil, but to produce heavy crops of large spears, it needs a well-drained, fertile seedbed that has been deeply prepared. It prefers a pH range of 6.0-6.8 but will grow on more alkaline soils if the major nutrients are in balance.

SOIL PREPARATION

More time and effort should be spent preparing your asparagus bed than any other part of the garden because it may last for 15 years or more and, once it is established, basic changes are difficult. Dig a 5 feet wide bed for 2 rows or a 3 feet wide bed for 1 row. Remove the topsoil and pile it to one side. Next, remove the subsoil and pile it separately. Dig the soil to a depth of about 18 inches. Mix rough organic matter and fertilizer in with the subsoil and return it to the trench. Blend finer organic matter and fertilizer with the topsoil and, if needed, limestone and return it to the trench, breaking up all clods to insure thorough mixing. Phosphorus and potash should be incorporated deeply.

Merely planting the bed and broadcasting fertilizer over the surface is inefficient because the nutrient ions such as phosphate and calcium will be "glued" to the surface layers and will move slowly down into the root zone. Nitrogen is the opposite. It will move down, through, and out of the soil. If your soil is heavy clay, thoroughly mix in gypsum at the manufacturer's recommended rates to help loosen the soil throughout the root zone. Preparing the soil as described here will make it bulkier, raising the level of the asparagus bed 6-8 inches above the surrounding soil, which will be helpful for good drainage and faster warm-up in the spring.

PLANTS PER PERSON/YIELD

Grow 10-15 plants for each person. Beginning the third year after planting, this should yield 150-175 spears, or about 12-15 pounds per season. This crop will need a 20-30 ft. row or a bed 5 feet wide by 10-15 feet in length.

SEED GERMINATION

To overcome the typically slow and uneven germination of asparagus seeds commercial growers make a disinfectant solution of 1 part bleach to 10 parts warm water and stir the seeds in it for 1 minute, no more. They rinse the seeds and soak them in 85-90°F water for 3-5 days. (Atop your water heater is a good place for maintaining the water heat.) After soaking, the seeds are dried enough to evaporate surface moisture. Disinfected and pre-soaked seeds will germinate rapidly and uniformly, even in cool soil, and rarely develop disease problems. Another method is to mix a packet of seeds with 1 cup of moist milled sphagnum moss in a plastic bag. Place it on top of a water heater or similar very warm spot to keep temperatures in the 85°F range. At the first signs of sprouting, plant the contents of the bag, moss and all. Milled sphagnum moss contains a natural substance that inhibits rotting and damping off organisms, so disinfecting with a bleach solution is not necessary although it does help soften the seed coat.

Soil Temp.	41°F.	50°F.	59°F.	68°F.
Days to Germ.	No Germ.	53 days	24 days	15 days

Soil Temp.	77°F.	86°F.	95°F.	104°F.
Days to Germ.	10 days	11 days	19 days	28 days

GROWING SEEDLINGS FOR TRANSPLANTING

To grow seedlings for transplanting, plant the pre-soaked or pre-germinated seeds out of doors in a protected area such as the corner of a fertile, well-weeded flower bed or in a shallow 5-7 gallon container. Space seeds 3 inches apart. If you crowd the seedlings, the roots may become entangled and damaged when you dig and shake out the seedlings for transplanting.

Asparagus seed is never planted directly where it is to remain. The seeds sprout slowly and seedlings remain so small the first season they can easily be lost among weeds. Direct seeding into a nursery row and transplanting the following spring makes more efficient use of garden space.

Seedlings can also be grown indoors. Fluorescent lights will be needed since the seeds will be planted in midwinter, when sunlight is weak and windowsills are cold. Soak or pre-sprout seeds (see SEED GERMINATION). Plant in styrofoam coffee cups with drainage holes punched in the bottom and filled with planter mix. Place the lights within 2 inches of the seeds for maximum light and radiant energy. Raise the lights to 6 inches above the seedlings when they have 4-6 leaves. Thin to the strongest seedlings and harden them for a week to acclimatize them to wind and cold. See the Planting Date chart for when to transplant.

TRANSPLANTING SEEDLINGS AND PLANTING DORMANT CROWNS

To transplant seedlings grown indoors, strip off the plastic cup or invert it and tap out the plant, preserving as much of the root ball as possible. If roots have circled the bottom, trim them off, or straighten them out before transplanting.

To dig the seedlings for transplanting, water the bed or container deeply about 3 days prior to digging. Use a spading fork to loosen and pry the seedlings, retaining as much of the root system as possible. When using containers, dump out the contents and shake the seedling loose. Prepare holes for planting before you dig the seedlings. Dig only a few at a time and transplant immediately, before the roots dry out. For seedlings grown outdoors water the bed or container deeply about 3 days prior to digging. Prepare holes for planting before you dig the seedlings. Loosen and pry the seedlings with a spading fork, keeping as much of the root system as possible.

If you have ordered dormant crowns, they will be shipped early, before they resume growth and develop spears that can be damaged in transplanting. Soak the crowns in water at room temperature for a day before planting and plant without delay.

When the asparagus bed has been prepared, plant dormant crowns or seedling transplants by digging a trench 6-8 inches deep down the center of a 3 foot wide bed, or by making 2 trenches down the shoulders of a 5 foot wide bed, and spacing crowns or seedlings 2 feet apart in rows, no closer. Research has shown that 2 feet x 5 feet spacing produces the best yields of large spears over the years. Spread out the roots in the trench, leading them away from the plant. The top of the plant should be 2-4 inches below the surface of the surrounding soil. Cover with pulverized soil and firm it lightly, but do not tamp it down.

Pour at least a quart of water over each plant to settle it and to start the uptake of water. Add a fertilizer starter solution to the water if you wish. When the plumes reach 8-12 inches in height, hoe soil up around the seedlings. Do this several times during the season as you cultivate. By fall, the original plant should be 6-8 inches below the soil level of the rest of the garden with the shoulders of the bed sloping away from it.

MULCHING AND CULTIVATING

You may hear folk advice to apply salt to asparagus beds. Research has proven salt is of little or no value in controlling weeds. Sodium in excess can displace ions of valuable nutrients on soil particles.

Mulch with organic matter except where snails, slugs, earwigs, and sowbugs are such a nuisance that clean culture is necessary. Mulch is generally added each year following the addition of organic or manufactured nitrogen fertilizer or, on sandy soils, a complete fertilizer. Add 6-8 inches of mulch around the plumes soon after spring harvest.

By fall the mulch will have decomposed and shrunken to a shallow layer that will not interfere with the next spring's harvest. The mulch should be coarse enough to admit oxygen and water. If you use leaves, they should be shredded. Peatmoss and packaged manure should not be used as mulch because both tend to seal over and shed water. Plastic sheet mulching is not practical since spears emerge over such a wide area.

Early spring, before any spears appear, is a good time to cultivate clean-culture beds. Keep cultivation shallow and use a tined tool. If you cut into any spears, discontinue cultivating. Don't let perennial weeds and grasses get started in the asparagus bed. Root them out at first sign. It is disheartening to have to replant an asparagus bed because it has become infested with quackgrass, nutgrass, bermudagrass, or wild morning glory.

WATERING

During the summer and fall restorative period when asparagus plumes are allowed to grow, the plants will need occasional deep waterings or rain. In a dry spring, water deeply once or twice during the spring harvest period. Let a water hose trickle around the plants for an hour or two, or sprinkler irrigate. When using sprinkler irrigation, set a glass under the spray pattern. When the glass has collected water to a depth of 4-6 inches, move the sprinkler. Either method should wet the soil to a depth of 20-30 inches. On deep, sandy soils it would be more beneficial to apply half as much water twice as often.

In dry western areas where winters are too mild to bring about a normal winter dormancy, a summer dormancy can be induced by letting the plants go dry. When a crop of plumes forms after cutting is over, stop watering for 45-60 days in midsummer. In late summer, break the dry dormancy with a heavy watering and supplementary feeding. Continue watering every 3-4 weeks if no rain falls.

CONTAINER GROWING

Asparagus is not suitable for container growing. The root system is so massive and so many plants are required to produce a meal for an average family that your containers are better used for other vegetables.

ENVIRONMENTAL PROBLEMS

The most common problems with asparagus are spindly spears and a gradual decline of plants. Spindly spears may be due to poor drainage, poor nutrition, shade from surrounding trees or shrubs, or a decline of vigor from such diseases as fusarium crown rot or asparagus rust. Aphids on the emerging spears and asparagus beetles on the maturing foliage are two of the most common insect problems.

HARVEST

Plants grown from crowns should go through two winters before spears are harvested. Plants from seeds should be given three winters. This allows the plants to develop a dense mat of thick roots for storing carbohydrates and nitrogenous matter, the "fuel" for creating long, thick spears. Harvest in the spring after new spears push through the soil. When spears are 7-8 inches long and thicker than a pencil, cut or snap them at or just below the soil level. The harvest season from established plantings is 6-8 weeks in the spring. All spears should be harvested. Stop harvesting when spears no longer grow beyond pencil thickness and let these develop foliage to nourish next year's crop.

STORAGE

Asparagus is one of the most perishable vegetables. Fresh asparagus may be refrigerated in an airtight wrap up to 3 days. Keep moist.

SEED STORAGE

When stored with a desiccant in a sealed container, asparagus seed should last for at least two years.

BEAN, SNAP -- BUSH OR POLE
(STRING BEAN)

Bean, Snap -- Bush Or Pole (String Bean)

Phaseolus vulgaris

Leguminosae (Pea Family)

Warm Season

Annual

Killed By Light Frosts

Full Sun

Ease Of Growth Rating, Bush Varieties -- 2; Pole -- 3

DESCRIPTION

Bush snap beans produce their pods on fast-growing, upright plants 14 to 18 inches tall by 12 inches wide. Pole snap beans differ in that they have 6-to-10-foot long, twining, climbing vines, are slightly later bearing and are more productive per foot of row, except in very cool areas. The price for this extra production is a extra labor in setting up and taking down arbors, tripods or netting but these offer another reward. You don't have to stoop over to pick the pods. Bush varieties will grow in all zones, pole varieties in all except zone 1.

Snap bean pods come in three colors, either green, yellow or wax, or purple. Pods come in a number of shapes, including round, oval and flat. Four of the most important considerations in snap beans are: good flavor, freedom from parchment and strings, resistance to important bean diseases, and the ability to set pods under both cool and warm weather conditions, leading to multiple pickings.

Snap beans are a native of the New World and were introduced to the colonists by the American Indians. Until the late 1800's, the pods of garden beans were stringy and full of parchment. They were either snapped and eaten after boiling, threaded on strings and dried for "leather britches beans", or grown to maturity and shelled to store as dried beans.

SITE

Snap beans should have full sun all day long. Although most bean diseases are not soil borne, it is wise to plant beans in a different spot every year.

SOIL

Beans prefer sandy loam or clay loam soils of pH 5.5 to 6.8, but can be grown on sandy soils if a fair amount of organic matter is present. Beans need minimum levels of phosphate and potash, and a moderate level of nitrogen, too much of which tends to stimulate excessive vegetative growth at the expense of pods.

SOIL PREPARATION

Light to moderate additions of organic matter are called for; this can be worked into the soil prior to planting. If wood ashes are used for potash and, to some extent, phosphate, they should be applied only lightly because of their caustic action and creation of alkalinity. If yours is a new garden, inoculate your bean seeds with nitrogen-fixing bacteria.

PLANTS PER PERSON/YIELD

Bush types—Grow 15 to 20 plants for each person to get 12 to 15 pounds of pods.

Pole types—Grow 12 to 15 plants for each person to get yield 12 to 15 pounds. Pole types yield over a longer period of time than the bush types.

SEED GERMINATION RATE

Soil Temp.	59°F.	68°F.	77°F.
Days to Germ.	16 days	11 days	8 days

Soil Temp.	86°F.	95°F.
Days to Germ.	8 days	6 days

DIRECT SEEDING

In most climates, direct seeding is the best way to start beans. Sow seeds ¾ to 1 inch deep, spacing bush varieties 1 to 2 inches apart. Space pole varieties in hills about 12 to 15 inches apart. For trellises sow 1 to 3 seeds in each hill. Use a non-crusting covering such as compost, sand or packaged potting mix. Another approach to avoid crusting is to open the furrow and water it thoroughly, sow the seeds and then cover with dry soil, avoiding any further overhead watering until the seeds sprout.

INOCULATION

Beans and other plants in the legume family have the ability to extract nitrogen from the air and fix it in their roots where it becomes a free source of fertilizer. Nitrogen-fixing bacteria are essential for this process. Some soils lack bacteria to do the job, especially new gardens. You can treat bean seeds by inoculation before sowing, thus assuring enough bacteria. On established gardens re-apply every few years. Purchase fresh powdered inoculant each year. It is usually available where you buy seeds.

TRANSPLANTING AND THINNING

Transplanting is possible but borders upon being impractical due to the cost and effort involved. Transplants can be started in peat pots 4 to 5 weeks before outdoor seed sowing time in the spring. This gives a jump on the season and helps reduce problems of poor germination caused by cold, damp soil. Set hardened transplants so the thick seed leaves are about one inch above soil level. Water thoroughly and protect with cloches or hot caps if frost threatens.

Thinning bush beans — in single rows, thin or transplant bush types to stand 2 to 3 inches apart. Rows should be 2 to 3 feet apart. Some gardeners alternate closely spaced double rows with foot-paths to make efficient use of space. You should then space plants a bit further apart, perhaps 6 to 7 inches. Space the rows about 7 inches apart. Wide band planting can be used, especially where garden space is at a premium, but it complicates weeding and hoeing. Bands of seeds should be between 18 and 24 inches wide, but never wider than can be straddled for easy picking. Space plants 4 to 6 inches apart in all directions.

Thinning pole beans — in single rows, thin or transplant pole beans 6 to 8 inches apart. Rows can be 2 to 4 feet apart. When pole beans are grown on teepees or similar supports, sow 3 to 4 seeds at each pole, thinning to 2 plants per location.

MULCHING & CULTIVATION

Beans are not ordinarily mulched because warm soil is desired for good production. During the warmest months, however, a light organic mulch can be used for weed control. Beans are shallow-rooted; don't cultivate deeply if you use clean culture.

WATERING

Beans need rain or irrigation every one to two weeks. Lack of water, depending on the water holding capacity of the soil during the blossom or pod-set period, can result in both blossom and pod drop, and in malformed, poor quality pods. Furrow irrigation is preferred because it does not wet the foliage. Apply one to two inches of water per irrigation.

CONTAINER GROWING

Bush beans can be grown in containers, but it is rather pointless. So many pods are needed to produce a meal of beans for an average family that numerous plants would be required. Pole or runner beans are more practical for container growing if you use a container of 10 or more gallons capacity. Use a somewhat heavier mixture than the typical soilless growing medium; fortify it with about fifty percent garden loam. Set the container by a fence and run strings to the top, drop stout strings from hooks in the ends of rafters or purchase a ready-made "tree" for supporting climbing vines.

ENVIRONMENTAL PROBLEMS

The most prevalent problem with garden beans is distorted plants and malformed pods caused by infestations of common bean mosaic virus. This disease can be brought in by feeding insects and can seriously damage or kill plants unless mosaic resistant varieties are planted.

Another common problem is failure to set pods due either to very cool weather or extremely hot weather, which interferes with pollen formation. Occasionally seen are malformed seedlings due to rough handling of the seeds or to badly crusted soil. The seed leaves can actually be broken off and only the slender tip will emerge or the tip may be broken off while the seed leaves remain. Either way, the seedlings will be delayed two to three weeks in maturity and are usually pulled out to make room for normal plants to develop. On heavy clay soils, hard rains soon after planting can seal in seedlings and prevent emergence. This can be lessened by shallow planting.

A common mistake in planting beans is to plant the seed too early, thus encouraging rotting or damping off in the soil.

HARVEST

Pick snap beans when the seeds inside are relatively small, before the sides of the pod begin to bulge noticeably. Most round and oval podded varieties should be about pencil diameter and 4 to 6 inches long at ideal harvest time. If harvest is delayed, the pods will become tough and less flavorful, and the seeds inside will become starchy. After the initial picking, beans will be ready for harvest every three to five days. Don't delay harvest, because pods left on the plant beyond full maturity will reduce production considerably.

Bean pods develop in trusses containing several fruit, but all are not ready for harvest at the same time. Pick beans by holding the stalk that attaches them to the plant in one hand while gently pulling the pod with the other hand. This way, younger, developing pods won't be disturbed. Compost the spent vines at the end of the season as they are a vegetable source of nitrogen.

STORAGE

Snap beans may be kept in the refrigerator in a moisture proof bag for one week. Do not wash before storing.

STORING LEFTOVER SEEDS

Seed placed in a sealed container with a desiccant and stored in a refrigerator should have a germination rate of at least 50% after two years, unless exposed to high heat and humidity before storage.

BEAN, LIMA, BUSH AND POLE

Beans, Lima, Bush And Pole

Phaseolus limensis **And** *P. limensis* **Var.** *limenanus*

Sieva Or Butter Bean Types, Bush Or Pole

Phaseolus lunatus

Leguminosae (Pea Family)

Warmth Loving Annual, Perennial In Parts Of Zone 7

Sensitive To Frost

Must Have Full Sun All Day

Ease Of Growth Rating, Bush -- 3; Pole -- 4

DESCRIPTION

Lima beans are not as popular as snap beans because the pods have to be shelled. The brisk sales of frozen lima beans are a better indicator of the popularity of the vegetable than their ranking in garden use. Lima beans are later bearing than snap beans and the foliage is immediately distinguishable, being longer, more pointed and somewhat more glossy. The pods are borne in clusters and, when the weather is warm and humid over an extended period, such large clusters are developed that you can quickly fill a basket with pods.

There really is a difference in "lima beans" and the "butter beans" of the South but it is slight. The original butter beans were high climbing, heat resistant pole types with small pods and small, thin seeds, snow white when dry. Some varieties were speckled, and had a strong flavor. The taste is much like the "lima beans" of the North but the texture is different, less mealy. Butter beans and limas will be lumped in some sections that follow.

Bush types are more popular than pole limas, especially in zones 2 and 3 where the extra earliness is needed. Bush types grow 12 to 14 inches tall and up to 18 inches across. Pole types can climb 10 to 12 feet; the butter beans are especially robust.

Pole limas usually outyield bush types by a wide margin. One gardener fed a small family with the produce from twelve plants run up and over a child's swing set. The vines also cast welcome shade.

Pole limas are preferred over pole butter beans in

zones 2 through 4 because they will set pods at slightly cooler temperatures and the pod fill is superior. Pole butter beans will bear through most of the summer in zones 5 through 7, going out of production only when temperatures are consistently 90°F. or warmer, and when the humidity has dropped. Bush butter beans are not as tolerant of heat nor as durable as the rugged pole types.

Both lima and butter beans are thought to be of South American origin, but lima beans were found growing in the wild in Florida where early indian plantations had been abandoned. They could have been a legacy of trading with other tribes who had ventured deep into Mexico.

SITE

Lima and butter beans must be grown in full sun, except in zones 6 and 7 where light afternoon shade can be beneficial in midsummer. (There is an art to situating a garden spot where it will receive afternoon sun but will not suffer from competing tree roots.) Do not plant lima beans in soil where you have grown limas during the past three years. Chop and compost the spent plants; they are high in nitrogen and will accelerate the decomposition of the compost heap.

SOIL

Limas and butter beans will grow reasonably well on a variety of garden soils but prefer sandy loam or well drained clay loam soils. They will not tolerate strongly acid soils, but will grow at a soil pH of 5.5 to 7.6. They are only moderate feeders. Phosphate and potash levels must be maintained to get a good set of well-filled pods.

SOIL PREPARATION

Limas benefit from a deeply prepared seedbed through which soil moisture can be pulled upwards by capillary action. A moderate amount of organic matter incorporated throughout the tilled soil will help to keep it open and well aerated. Incorporating lime (where needed) and phosphorus, rather than dressing it on the surface, is beneficial because this places calcium and phosphorus where feeder roots can reach the particles. These two minerals stay where you place them in the soil and don't gravitate down. (They are exceptions; nitrate nitrogen and, to a lesser extent, potash, will move down through the soil when dissolved in water).

Raised beds or ridged-up rows are recommended for limas in zones 2 through 4 where the improved drainage and greater absorption of solar energy will lead to earlier crops. Planting on the flat is common elsewhere except where raised beds or rows expedite drainage of heavy soil or where furrow irrigation is practiced.

PLANTS PER PERSON/YIELD

Pole types bear over a longer period and provide harvests throughout the summer and, except in zones 6 and 7, where they may burn out, into the early fall season. Grow 10 to 12 plants of pole types for each person. In the North this should yield 6 to 8 pounds of shelled beans — in the South or the warm West, 8 to 10 lbs. Grow 12 to 15 plants of bush types for each person. This should yield 6 to 8 pounds of shelled beans in the North — more in the South or warm West because of the extended harvest period.

SEED GERMINATION

Limas require warm soil to germinate quickly and well. While the chart below indicates they will sprout over a wide range of soil temperatures, in cool soil the seeds may decay or sprouts may die before emerging.

Soil Temp.	59°F.	68°F.	77°F.
Days to Sprout.	30 days	18 days	6 days

Soil Temp.	86°F.	96°F.
Days to Sprout.	6 days	7 days

The seeds of lima and butter beans are easily damaged by rough handling in the seed fields, in cleaning and packaging, and in shipment. While home-saved seeds might germinate 90 to 95%, purchased seeds will probably germinate 75%. One sure way to realize the maximum growth potential of lima seeds is to cover them with sand to trap solar heat and to keep crusts of hard soil from forming over the seeds. Some gardeners turn the seeds of large-seeded limas on edge when planting but this is not necessary when the soil is amended with sand, or is naturally sandy.

DIRECT SEEDING

Lima bean seeds require a soil temperature of at least 65°F. to germinate well; delay planting until 10 to 14 days after the last spring frost date. Limas are normally sown 1 to 1½ inches deep and 3 to 4 inches apart. In heavy clay soils either cover seeds with sand or reduce the planting depth to ½ inches. On dry, sandy soils, increase the planting depth to 2 inches. Pole beans are usually planted in "hills", groups of 2 or 3 seeds per support, to reduce the chance of skips. The use of an inoculum is recommended.

TRANSPLANTING AND THINNING

You can get the jump on the season by sowing lima seeds indoors in peat pots four weeks prior to the outdoor direct seeding dates. Even this extra effort won't let you grow limas in zone 1, where not enough solar energy is accumulated to mature a crop before frost. However, in a warm year in zone 2, transplanting could add a week or two of production. Beans are hypersensitive about root disruption;

grow the seeds in peat pots and transplant them when the plants are quite small, before the taproot grows out through the bottom of the pot. Space bush types 3 to 5 inches apart in rows spaced 2½ to 3 inches apart, and pole types 6 to 9 inches apart for trellises. For teepees, leave two plants per pole in the event one is damaged or killed.

MULCHING AND CULTIVATING

Black plastic is sometimes used to hasten the maturity of bush and pole types in zones 2 and 3. It isn't necessary further south. Organic mulches are not ordinarily applied in zones 2 and 3 nor on heavy soils anywhere because they tend to keep the soil too moist. They work best on lightweight soils which tend to run drier than lima beans prefer.

Cultivate the soil by scraping off weeds with a sharp hoe; deep hoeing can damage roots seriously, set back maturity and diminish yields.

SUPPORTS AND STRUCTURES

Gardeners who erect weak or short supports for pole varieties of lima or butter beans are more to be pitied than censored. It is virtually impossible to replace or lengthen collapsed or inadequate structures once they are loaded with bean vines.

The problem usually starts with trying to build supports with materials found around the house or yard. The typical suburban home doesn't have a thicket of tall, straight saplings or bamboo cane to cut for pole arbors, nor wire heavy enough for stringers between posts.

The best bean arbors are made with 8' poles, bamboo or 2 x 2 slats painted with copper napthenate as a wood preservative. These should be stuck into moist soil to a depth of 12" to keep the structures stable during windstorms. Three or four poles can be lashed together near the top to make a teepee or pairs can be lashed and connected with a top runner down the length of row.

More permanent supports can be made of 4" x 4" x 8' landscape timbers placed 6 to 8' apart for posts and sunk at least 2' into the ground. These are connected with top and bottom wires of 10 gauge or heavier. Strong vertical strings or twine are laced between the wires to train vines.

Posts of 8' length are too short for most bean varieties. When you set them to a depth of 2' or more in the soil, the uprights are too short for the long runners of lima or butter beans. They hit the top and cascade down, making a tangle that is difficult to pick.

WATERING

Lima beans dislike droughty soil. They will survive dry periods, but you can see skips in the set of blossoms and pods. Hot weather, too, is hard on the plants, but the effects can be lessened by adding enough moisture to the soil to avoid stressing them. Yield is directly related to uniform soil moisture. If no rain for 10-14 days, apply 2 to 3 inches of water, preferably by furrow irrigation.

CONTAINER GROWING

Pole lima or butter beans make practical container plants and, if kept free of insects and run up strong strings, can be most attractive. Use them to make shaded arbors for your patio.

ENVIRONMENTAL PROBLEMS

In the South and certain parts of the West, lima and butter beans are affected by verticillium and fusarium soil diseases and by nematodes, a worm-like, microscopic soil-dwelling creature. Spider mites and Mexican bean beetles can become a serious problem late in the season, especially during dry weather. Generally, lima beans are more free of diseases than snap beans, but they are less tolerant to weather stresses, except certain varieties of butter beans.

HARVEST

Most gardeners prefer to shell green limas for fesh use rather than to let them dry for storage. Harvest when the seeds enlarge to fill the cavities inside the pod. Try to pick the pods before they turn yellow. However, a few mature seeds from yellowing pods never hurt a 'mess' of lima beans. Depending on the variety, the seeds may be either creamy, greenish-white or speckled. The entire pod is harvested by first feeling it for plumpness, then carefully pulling it from the plant. Remove the seeds by shelling (splitting open) the pod, a time consuming ceremony that can be performed automatically while watching TV, or the neighbors from a front porch rocker.

If you want limas seeds for dry storage, leave the pods on the plant until they turn brown and the seed rattle inside. Pull the dried pods before they split open, because they would then drop their seeds. Shell out the seeds and place them in a dry, airy location until the seed coat hardens. Drop a dried hot pepper pod or two into the jar before sealing, to repel weevils and worms.

STORAGE

Lima Beans can be kept fresh in the refrigerator in an air tight container for 5 days.

SEED STORAGE

Seeds placed in a sealed container with a desiccant and stored in a refrigerator should germinate about 50% after two years.

BEANS, GREEN SHELL AND DRY

Bean, Green Shell And Dry

Baking Bean, Soup Bean (Horticultural, Navy, Great Northern, Pinto, Kidney, Black, Pink Tepary)

Phaseolus vulgaris

Leguminosae (Pea Family)

Warm Season

Annual

Killed By Light Frosts

Full Sun

Ease Of Growth Rating -- 2

DESCRIPTION

Bean pods can be shelled for seeds at both green and dry stages. Technically, any bean variety can be shelled, including the standard garden snap bean varieties developed for their tender pods. However, the specialized shelling varieties cook to a better texture and a more attractive liquor color.

The varieties developed exclusively for dry beans usually have bushy plants, although certain varieties have short to medium length runners and are sometimes referred to as half runner types. These are extremely productive, rather late maturing beans, and modern varieties have been bred to resist diseases and tolerate weather stresses. Most varieties of dry beans have larger plants than bush snap beans; those with runners are self-supporting but can be grown among corn plants where the short runners twine around the stalks for support.

Some varieties of green shell beans can be used for snap beans when the pods are young and for shelling when the seeds are full size but still green and tender. However, when the pods are allowed to dry on the vine, they produce a mealy "dry bean." Certain varieties of green shell beans have such fibrous pods, even at early stages, that they are always shelled from the pods for eating. Green shell beans make an acceptable substitute for lima beans in areas too cool for good lima production.

Dried beans are an ancient food crop originating in the New World and prized as a concentrated food source that can be stored for a long time. Dry bean seeds have been found in cave deposits dated at 4,900 BC in Mexico and at 500 BC in the USA. Tepary beans, *Phaseolus acutifolius* var. *latifolius,* were taken from the wild for improvement because of their tolerance for dry soil, hot weather, and slightly more alkalinity than *P. vulgaris*. Green shell beans were, at first, simply taken from immature pods of dry storage types. In the 19th century special varieties, called "Horticultural Beans," were developed for green shell use, and used as snap beans when young.

SITE

Shell beans and dry beans should be grown in full sun. Although many bean diseases are not soil borne, it is, nevertheless, a wise precaution to plant beans in a different area every year. Since shell and dry beans are not space efficient, only gardeners with fairly large food gardens grow them in substantial amounts. Dry beans are usually inexpensive, and it is hard to justify diverting limited space from more profitable crops.

SOIL

Beans are only moderately tolerant to soil acidity and will grow best at pH levels of 5.5-6.8. They prefer well drained sandy loam or clay loam soils. However, they can be grown fairly well on sandy soils containing 2-3% of organic matter, an amount sufficient to give a dark tinge to the soil. They need a moderate level of phosphate and potash, but not a great deal of nitrogen which tends to stimulate excessive vegetative growth with no increase in pod production. Beans are valued by organic gardeners as a soil restorer, because of the nitrogen fixed from the air by bacteria on the roots and for the high nitrogen content of the foliage for green manure.

SOIL PREPARATION

Work light to moderate amounts of organic matter into the soil prior to planting. Except on sandy soil, beans planted in late spring will benefit from raised beds or ridged rows which improve drainage and warm the soil faster.

PLANTS PER PERSON/YIELD

Grow at least 20-30 plants for each person for green shell use and for storage to get 3-5 pounds of green shell beans or 2-4 pounds of dry beans.

SEED GERMINATION

Seeds will germinate at soil temperatures of 60-95°F, but the optimum is 80°F. Some varieties of shell beans have a fairly high content of "hard seeds." This is a protective genetic mechanism which can delay germination for a week or two. Therefore, don't be surprised if certain varieties of dry beans germinate over an extended period of time. Bean seeds should germinate 75-80% on the average if fairly fresh and if not damaged in harvest or shipping. Bean seeds may rot if planted in soil cooler than 60°F and will emerge slowly when the soil starts to warm. Under such stress conditions, losses before and after emergence can be severe.

DIRECT SEEDING

Plant after the soil has warmed to at least 60°F and all danger of frost is past. A common mistake is to sow the seed too early, which encourages rotting or damping off in cool, wet soil. Green shell or dry beans should be direct seeded ¾-1 inch deep. Space bush types 1 inch apart and types with short runners 2 inches apart. However, if the row has gaps you can move surplus seedlings just after they have emerged, by first digging a hole to accept the seedling, then moving it with a trowel-full of soil to preserve as much as possible of the root system. Water the transplanted seedling at once. Most transplants will take, but maturity may be delayed because of damage to the feeder root system. Cover seeds with sand, vermiculite, or finely sifted compost unless your soil is sandy or non-crusting. If yours is a new garden, inoculate bean seeds with nitrogen-fixing bacteria.

TRANSPLANTING AND THINNING

Beans are almost always direct seeded. Transplanting damages the fragile root systems, so that any advantage gained by starting early indoors would be lost during transplanting. Space rows of bush types 2 feet apart and half-runner varieties 3 feet apart, or interplant them with sweet corn, popcorn, or field corn.

Bush types in single rows should be thinned to stand 3-5 inches apart and types with short runners to stand 6-8 inches apart. When planted close together in double rows or wide bands, plants should be thinned to stand 6-7 inches apart.

MULCHING AND CULTIVATION

Ordinarily, beans are not mulched since warm soil is desired for good production. However, if your soil drains well, an organic mulch can be applied for weed control. It should be only 2-3 inches deep so that the soil is not kept too cool or moist.

Avoid deep cultivation. Production can be seriously hurt by cultivating beans with a tiller. Hoeing should be shallow; scraping is preferable.

WATERING

Green shell varieties tolerate dry soil about as well as snap beans, while dry beans, especially the vigorous types with short runners, and tepary beans are more drought resistant. All beans, however, need a steady supply of soil moisture during the blossom and pod set period. Furrow-irrigate beans to avoid encouraging foliage diseases. Apply 1-2 inches of water per irrigation to wet the soil to a depth of 5-10 inches. When there is no rain, sandy soils may have to be watered as often as every 4-5 days, but clay soils can generally go for 7-10 days. Loamy soils come in between these extremes in water needs.

CONTAINER GROWING

Bush green shell or dry beans do not make efficient use of container space. Also, while beans will grow in containers, the vines become unattractive when under the stress of a heavy crop of pods, especially when they are drying. Pole varieties of green shell beans can be grown in containers and strung up supports.

ENVIRONMENTAL PROBLEMS

The most prevalent problem with green shell and dry beans is distorted plants and malformed pods caused by infections of common bean virus, brought in by feeding insects. It can seriously damage or kill plants. Another common problem is failure to set pods, due to very cool weather or extremely hot weather, both of which can damage the pollen. Occasionally, seedlings are malformed due to rough handling or by breaking through crusted soil. The seed leaves can snap off and only the slender tip will emerge, or the tip may break from crusted soil and only the seed leaves will remain. Either way, the seedlings will be delayed 2-3 weeks in maturity, and are usually pulled out to make room for normal plants to develop.

HARVESTING

Harvest beans for green shell use when the seeds are plump and the pod is still moist. Mature green seeds, sometimes called "shellies" or "shuckies," shell out easily and quickly from pods. To harvest dry beans, wait until the pods and plants dry completely. Pull the pods from the plant and spread them to dry on a cloth or screen away from moisture. Place only one layer at a time on the drying cloth. When the pods are completely dry, thresh the seeds by rubbing the dry pods. Separate the chaff from the seeds by winnowing. On a breezy day, toss bean seeds from a shallow basket into the air and catch them, letting the breeze blow the chaff away.

STORAGE

Dry beans can be kept in a closed container for 1 year or more. Pick pods when they are mature. Spread in a warm, dry place until pods are thoroughly dry. Shell beans. To protect them from insects, heat the beans in a 180°F oven for 15 minutes. Turn off the heat and leave the beans in the oven for 1 hour. Cool. Store in a cool, dry place in a covered container.

Some gardeners prefer to mix a few dried hot peppers with dry beans for storage; they are an effective insect repellent.

STORING LEFTOVER SEEDS

When stored as recommended, bean seeds should germinate at least 50% after 2 years.

Beet

Beetroot, Garden Beet, Red Beet

Beta vulgaris

Chenopodiaceae (Goosefoot Family)

Cool Season

Biennial, Grown As An Annual

Winter Hardy In Most Climates

Full Sun Or Partial Shade

Ease Of Growth Rating -- 4

DESCRIPTION

Beets are a popular vegetable, used for both roots and edible tops. They are a space-efficient crop and yield much good food from a small area. Modern beet varieties will keep in the soil for weeks with no loss of flavor if you cannot harvest them at peak maturity.

Beet varieties can be divided roughly into two classes: the globe type with a small amount of "shoulder" protruding above the surface of the soil and the cylindrical type with about one-third of the root above the soil level. While all beet varieties have edible roots, some were developed to produce heavy crops of succulent tops as well. Greens from these dual purpose varieties are somewhat larger and more tender than those from beets developed principally for roots. Beet greens have a distinctive taste, different from that of chard.

F_1 hybrid beets are now appearing in home garden catalogs. The seeds are more expensive than those of open pollinated varieties. Some offer home gardeners beets with greater uniformity, additional vigor, and higher quality than older varieties. However, open pollinated beets have been under improvement for so long that, except for a moderate degree of variation in root size and shape, they are consistently high in quality.

Beets are fairly frost-hardy, but will not endure heat in the seedling stages. Therefore, midsummer is usually skipped as a planting time except in zones 1 and 2 and the northern portion of zone 3. Spring beet crops in zones 6 and 7 are timed to be harvested before the hottest part of summer, when hot soil and water stresses diminish their quality.

Beets are native to Europe, North Africa, and West Asia, where they grow wild as coarse, thick-rooted perennials. In German literature the first mention of beets was in 1558; in England, 1576. In 1806 the McMahon catalog mentioned beets being grown in the gardens of America, but it is not known definitely when seeds were first brought over. Swiss chard, sugar beets, and stock beets or mangel-wurzels all belong to the same species as garden beets. DeCandolle states that although the ancients knew about the beet, they did not cultivate it until the third century A.D. German and French seed companies became interested in improving beets about the year 1800; since that time many classes and varieties have been developed.

SITE

Beets are adaptable but prefer full sun, except in the hottest parts of the South and warm West where they benefit from light afternoon shade from midsummer through fall.

SOIL

Beets will tolerate a wide range of soil conditions, but grow best on sandy loam or fast-draining clay loam soils. On fertile soils with a consistently good supply of moisture and plant nutrients, beets will develop quickly and will be almost free on "zoning," the white rings which are sometimes evident when you cut the roots. Zoning is more evident when the weather ranges from quite dry to moist or when the plants have been stressed for extended periods. Beets are only slightly tolerant of acid soils and grow best at a pH range of 6.0-6.8. They will thrive on "limey" soils with a pH level of up to 7.5, but growing on such alkaline soils increases the risk that beets will develop a disease called "scab." You can decrease the risk of scab by incorporating organic matter into the soil.

SOIL PREPARATION

Beets like plenty of organic matter worked into the soil, but it should be screened. Rough or coarse organic matter in the soil may cause beet roots to become hairy or misshapen. Use well rotted, screened compost, peat moss or sifted forest compost. Except on sandy, fast draining soils, beets grow better if you raise the beds somewhat for good drainage. Beets are difficult to grow on ridged-up rows because the soil tends to erode away from the roots.

PLANTS PER PERSON/YIELD

Grow 40-50 plants for each person to get 8-10 pounds of roots plus a few pickings of tops.

SEED GERMINATION

Most varieties of beets produce seeds in clusters called "seed ball." Two or 3 seedlings may grow from a single seed ball. The seeds germinate and grow slowly in cool soil; little is gained by extra-early spring planting. The optimum soil temperature for germination is 77-86°F. Two techniques for speeding germination are pre-soaking and pre-sprouting. To pre-soak, wash seeds in a mild solution of dishwashing detergent, rinse several times, and soak in lukewarm water for 3 hours to soften the corky seedcoat. Drain before planting. To pre-sprout, mix a packet of seeds with a cup of moist peat moss or

BEETS

milled sphagnum moss in a plastic bag. Store in a dark area at 65-75°F until the first sprouts appear; then scatter the contents of the bag down a shallow furrow and cover with ½″ of sand or vermiculite. Sprinkle once or twice a day until seedlings emerge.

Soil Temp.	41°F.	50°F.	59°F.	68°F.
Days to Germ.	42 days	19 days	10 days	6 days
Soil Temp.	77°F.	86°F.	95°F.	104°F.
Days to Germ.	5 days	5 days	5 days	No Germ.

DIRECT SEEDING

To speed germination, pre-soak seeds, (see Seed Germination.) Space the seed balls about ¾ inches apart and cover ¼-½ inch deep. If you pre-sprouted seeds (see Seed Germination), scatter the contents of the sprouting bag down a shallow furrow covering ½ inch deep with sand or vermiculite and sprinkle once or twice a day until seedlings emerge. Space single rows 12 inches, 18 inches for leaf varieties apart.

Beets are a natural for wide band planting. Prepare a seedbed the width of a garden rake and sow seeds about 1 inch apart in all directions. When germinating beet seeds in summer for an autumn crop, sow pre-soaked seeds at least ½ inch deep. Cover the seedbed with burlap to reduce drying. Water through the burlap and remove it when the first seedlings appear.

TRANSPLANTING AND THINNING

In zones 1 and 2 early spring beets can be started indoors 6-8 weeks prior to the transplant date and transplanted 2-3 inches apart as small seedlings. Cover seeds ¼ inch deep and grow in strong sunlight or under intense light from fluorescent lamps.

There is no need to thin beets which are sown the proper distance apart. Even though 2 or 3 seedlings may come up from a single seed ball the strongest seedling will dominate the weaker and will soon take over. With "monogerm" beet varieties (those bred to produce only 1 seed per seed ball), more exact spacing of plants is possible.

MULCHING AND CULTIVATION

Beets respond favorably to mulching with 2-3 inches of coarse organic matter such as rotting straw, pine needles or hay, partially or fully composted sawdust, or composted refuse from last year's garden. In some garden literature you will find cautions against mulching root crops. Such warnings are the result of bad experiences from applying a deep mulch on heavy, cold soils too soon in the spring, before they have warmed and dried. In addition to their beneficial effects on soil moisture and temperature, mulches keep beet foliage clean and produce roots free from sun-roughened shoulders. This is particularly the case with cylindrical beets,

which should be mulched with 3-4 inches of organic litter to prevent roughening of the shoulders that protrude above the ground. For beets which are planted in late summer for fall harvest mulching will contain the soil warmth, prolong growth, and protect the roots against the first heavy freezes.

In clean cultivation, soil is scraped up around the beet roots gradually so that the shoulders are covered at maturity. This shallow cultivation not only hills up and covers beet shoulders but also reduces weeds. Some hand pulling of weeds between plants and rows may be necessary.

WATERING

Regular watering or frequent rains encourage the best quality roots, sustaining steady, rapid growth and preventing checks in growth that result in "zoning." Sprinkler, furrow, and drip irrigation work equally well for beets. If no rain falls for 7-10 days, apply 2-4 inches of water to moisten the soil to a depth of 10-20 inches.

CONTAINER GROWING

Beets are one of the preferred crops for container growing. Deep containers are neither necessary nor efficient for beets. Shallow, 3 to 7 gallon containers are preferred. Beets can be closely spaced, 2-3 inches apart, and eaten as they are thinned. Beet seeds germinate quickly and grow rapidly in the fast-draining, highly organic, artificial growing media customarily used in containers. Consider planting more than one variety of beets in a large container; for example, a variety with bright green foliage plus one with reddish foliage for contrast.

ENVIRONMENTAL PROBLEMS

Usually, the first problem new gardeners encounter is poor seed germination, most often traceable to how the seeds are sown rather than to the seeds themselves. Germination can be improved either by pre-sprouting or pre-soaking seeds (see Seed Germination). In the garden cover sprouts or pre-soaked seeds ¼ inch deep with sand, vermiculite, or compost, but not with heavy garden soil which tends to crust and impede germination.

Woody or white-zoned roots are often seen in beets which are stressed by weather. In the southern portion of zone 3 and zones 4, 5, 6, and 7 the best way to avoid zoning is to plant beet seeds either quite early in the spring or after the hottest summer days are over. Avoiding midsummer and its extreme weather stresses in the middle and southern zones will yield higher quality beets. Beet tops that are overly red, lacking green color, are usually caused by malnourishment resulting from growth stress. This occurs most often in poor, dry soil. Also, red coloration in the leaves is intensified by cool temperatures.

HARVESTING

Pull the entire plant when the root has swollen to golf-ball size, or, if it is a cylindrical variety, when the roots are 4 inches in length. Then the root is at its tender, flavorful best and the tops are perfect for cooking as greens. Beets can be left in the ground to reach a diameter of 3 inches or more but at this size they often lose flavor and tenderness. To reduce bleeding during cooking, leave 1 inch of leaf stalks attached to root, and do not cut the root before cooking.

STORAGE

Before storing beets in the refrigerator, remove all leafy tops and stems to within 1 inch of the beet. Refrigerate in perforated moisture proof bag up to 2 weeks. Beets can be kept in a root storage area at 32-40° F for 1-3 months. High humidity is necessary to prevent shriveling. Either bury the beets with ½ inch stems in moist sand or place in plastic bags with several small holes.

To store beet greens: Wash, drain, and refrigerate in a moisture proof bag 1-2 days.

STORING LEFTOVER SEEDS

When stored as recommended, beet seeds should germinate more than 50% after 2-3 years of storage.

BROCCOLI

Broccoli, (Sprouting Broccoli, Italian Sprouting Broccoli, Calabrese)

Brassica oleracea (Botrytris Group) Modern, Large-Headed Broccoli

Brassica oleracea (Italica Group) Italian Sprouting Broccoli

Brassica rapa (Ruvo Group) Raab, Rapa, De Rapa Or Turnip Broccoli Cruciferae (Mustard Family) A "Cole Crop"

Cool Season Annual

Killed By Prolonged Freezing Weather

Full Sun Or Partial Shade

Ease Of Growth Rating, From Transplants You Start From Seeds -- 4; From Direct Seeding -- 3

DESCRIPTION

With the possible exception of snap peas, broccoli has grown in popularity more than any other vegetable in recent years. It was little known in this country until the heavy influx of Southern European immigrants around the turn of the century. Its current popularity is due to its introduction as a frozen food product after World War II to the recent availability of early, large-headed, bolting resistant varieties and to a growing appreciation of its high food value.

Much confusion exists in the broccoli 'family tree'. Originally, three types were grown in Europe: the late maturing "cauliflower-broccoli" type which resembled a green or purple-headed cauliflower, the early "Italian green sprouting" and Raab or Rapa, a distinct species.

Plant breeders selected from and crossed the sprouting and heading types to develop an early broccoli with large central heads, borne on erect plants, 12 to 18″ high. Unfortunately, they continued to call the new vegetable "sprouting broccoli".

Up-to-date catalogs cut through the confusion by using the term "sprouting broccoli" only to identify such old-fashioned varieties as Calabrese that do not form a large central head. This long-bearing type is a favorite of Southern Europeans and Orientals because its many small buds extend the harvest season and give edible tip leaves as well as buds.

Modern broccoli is so early that plants can form large central heads up to 2 lbs. in weight, or more, in the late days of spring, before summer heat and long days force the clustered buds to burst into flower.

Raab or Rapa broccoli looks more like mustard than broccoli and has long, branching stems with small, dime-sized bud clusters. It is so unlike conventional broccoli that some catalogs emphasize that this broccoli is an ethnic specialty.

Broccoli is a member of the Cruciferae family, so named because the flowers have four petals in the form of a cross. Found in the wild in Europe.

SITE

Broccoli is one of the few vegetables that will tolerate light shade, but be aware that any amount of shade on spring broccoli will delay maturity and make the plants grow open and lanky, with smaller heads. In zones 6 & 7, broccoli direct-seeded in late summer is often given light shade from the afternoon sun. To reduce the chance of soil-borne diseases, choose a site where no other mustard family member has been grown during the past three years.

SOIL

Broccoli is only slightly tolerant of soil acidity and grows best at pH levels of 6.0 to 6.8. A pH level below 6.5 can increase susceptibility to clubroot disease. In general, most moderately drained to well drained soils can produce good broccoli if supplied with plenty of plant food and water. Broccoli needs to grow and develop rapidly, free of growth checks.

This is especially true of spring broccoli which needs to "make" before hot weather. In deep, sandy soils, unless they are fortified with organic matter, watered regularly between rains, and supplemented with plant nutrients, broccoli may not head properly. Broccoli is a heavy feeder and requires good levels of phosphate and potash as well as nitrogen.

SOIL PREPARATION

Spring broccoli plants are usually set in the garden so early in the spring that preparing the soil the previous fall is preferred. The root system of broccoli doesn't go deep; tillage to spade depth is sufficient. Broccoli planted in late summer grows well if it follows a vegetable for which the soil was heavily fortified with organic matter.

PLANTS PER PERSON/YIELD

Grow 6 to 8 plants of standard broccoli for each person to get 8 to 10 lbs. of heads. Laterals or side shoots will be a bonus. Grow 8 to 12 plants of rapa; it yields less per plant per picking but gives multiple harvests.

SEED GERMINATION

Germination is easy because broccoli seeds are relatively large and fast sprouting. They will come up in only 3 to 7 days at 75°F soil temperature. The Federal Minimum Standard for germination is 75% but most seed lots will germinate 80-85%.

PLANTING DATES

Increasingly, gardeners are planting broccoli for fall harvest in place of, or in addition to, spring broccoli. Direct seeding is commonplace in late summer; the seeds come up fast in the warm soil.

Take note that broccoli seedlings are slightly less cold tolerant than cabbage. If you set them in the garden extremely early, even after hardening off, you could lose them to a hard frost.

DIRECT SEEDING

Transplants are usually preferred for spring crops, direct-seeding for fall crops. Plant 2 or 3 seeds together in a group. Space the groups 9 to 12 inches apart. Plant seeds ½" deep. This should guarantee one strong seedling per location. Make rows 18 to 24" apart, with the close spacing reserved for fertile, improved soils. If you prefer to run your tiller between rows for weeding, set the row spacing at 30" with 7 to 9" between groups of seeds.

TRANSPLANTING & THINNING

Start broccoli transplants from seeds indoors 4 to 6 weeks before the transplant date shown in the chart. It is difficult to grow good sturdy seedlings indoors without fluorescent lights because, in early spring, the days are short and sunlight through windows is weak. Plant 2 or 3 seeds per peat pot and cover ¼" deep. Use potting soil rather than garden soil to reduce the chance of damping off disease.

Indoor-grown seedlings are tender and can be harmed by cold if not hardened off or protected with Hotkaps or other protective devices for about a week after transplanting.

Depth of transplanting should be to the level of the first set of leaf stems, no deeper. This should position the plant ½ to 1" deeper in the soil than it grew in the container. This is somewhat deeper than for other transplanted vegetables but works to anchor the plants firmly erect and to keep them from toppling over and growing up into a gooseneck.

If you purchase transplants, avoid broccoli that shows a central flower bud or that is purple and wiry looking. Also avoid very large transplants; 1/3 the large surface area of leaves will need to be trimmed off to keep them from wilting when set in the garden.

Optimum spacing between broccoli plants in the row ranges from 7" for compact varieties on fertile soil to 18" for large-frame varieties on poor soil or where irrigation water can't be supplied. Wider spacing between rows usually goes hand in hand with close spacing within the row.

MULCHING AND CULTIVATION

Plastic mulch works well for spring broccoli but organic mulches are generally preferred for fall crops because they keep the soil cool and moist during the critical early growth stages. On spring crops the plastic mulch should not be applied until just before the plants are set in, to give the soil as much time as possible to dry out from melting snow or winter rains. Under the plastic mulch, nitrogen from decomposing organic matter will convert more rapidly to the nitrate form that can be utilized by plants. The solar heat trapped by the plastic stimulates beneficial bacteria to early activity.

On fall broccoli crops an organic mulch of straw, hay, half-rotted leaves or dried grass clippings over cardboard will work to insulate the soil and to conserve moisture. A 2" layer can be applied while plants are small and deepened to 6 to 8" after plants have gained sufficient height.

Weeds may be a problem in direct-seeded broccoli because weed seeds will sprout thickly in the warm, moist soil. Scrape them out from between and around the seedlings while they are small, using a slender, sharp-pointed knife. Later, hoe to control weeds but keep it shallow to avoid injuring the shallow network of feeder roots.

SUPPORTING STRUCTURES

Only in very windy areas or when it is winter-grown during California's rainy season does broccoli need the support of stakes. Mature or nearly mature plants may lean over after a severe windstorm but can be righted the next day.

WATERING

The large leaves of broccoli evaporate (transpire) a lot of water and the crop needs either rain or irrigation weekly. Wet the soil to a depth of 10" to 15" by applying 2" to 3" of water. Either sprinkler, drip, or furrow irrigation is satisfactory. At the first signs of mildew, discontinue sprinkling.

CONTAINER GROWING

Decorative, quick-growing, early bearing broccoli is a favorite for container production because it gives rather high returns for the space it occupies. Single plants of compact varieties will fit in 3-gallon containers but a better arrangement is 2 plants per 5-gallon bucket or 3 per 7-gallon tub. Spring crops can be direct-seeded in containers and protected with clear plastic which traps heat and shelters plants from frost.

ENVIRONMENTAL PROBLEMS

Rushing the planting season, especially in zones 1 & 2, does little good and can result in loss of the crop due to late freezes. Broccoli needs cool weather during the three or four weeks leading up to harvest. If cool weather is present, the plants have a good chance to form large central heads and then hold for a few days if not harvested immediately. The closer you can come to planting at the recommended dates, the better your chances for a good crop, especially with spring broccoli.

Beginning gardeners almost always plant broccoli later in the spring than they should, and the plants either don't bud at all, or they burst into flower when the plants and heads are still small. The latter is mostly caused by using transplants which are too old or too thoroughly hardened to regrow rapidly when transplanted. Their growth is checked, but not the flowering instinct. Thus, tiny heads on small plants, a condition known as "buttoning".

Broccoli plants can fall over if set in too shallow, or have their growth checked if set too deep in heavy soil and the soil firmed too vigorously around the root ball. Small plants can be set in at the depth they grew in the container and will not fall over. Top heavy large plants can be set 1", no more, below the depth they grew in the container.

Mortality is a major problem when transplanting field-grown plants that have not been conditioned or that are set in the garden on a windy day and not watered at once. In zones 5, 6 & 7, bunched, field-grown plants of cole crops are often sold. These large, bare-rooted plants are shipped to retailers in crates. Upon purchase the bunches should be opened and the plants stood in a bucket containing a few inches of water to which a little clay soil has been added, a process known as "mudding." Before planting, about 1/3 of the top foliage should be trimmed off to reduce water loss, and the holes dug and filled with water before plants are set in. Handled thusly, the plants should recover within a few days and resume growing.

HARVESTING

For fresh use, harvesting can begin when heads are only half grown. Harvesting of central heads should be completed while the buds are still tight and green, before heads loosen and the buds begin to break and show yellow. This means you should be prepared to freeze the excess.

If you delay harvesting broccoli until the flower buds open, it can still be cooked and eaten fresh, but the flavor will be strong. Such a delay will weaken the plant and reduce the crop of side shoots which form later.

To determine where to cut the central head, test the stem below the head with your thumbnail. Cut where the skin begins to become tough, or even 2 to 3 inches lower, if you will peel off the tough skin.

Side shoots, which develop in the leaf axils after the main head is harvested, are usually larger on fall crops. They should be harvested when the buds are about 1" across and still tight and green. In zones 4 through 7, where summer often comes on early and hot, side shoot production can be restricted in size and quantity.

Heads of sprouting broccoli are numerous but small, and stems are long. Harvest 2 or 3 times weekly during the cutting period. Buds can reach and pass the best eating stage quickly. In practice, you will find yourself removing all but tiny, dime-sized, immature buds each time you harvest and that your produce will contain buds ranging from tight to half open. Snap or cut off stems at the point where the skin begins to feel woody. With sprouting broccoli and raab, there is no clear break between the formation of primary and secondary buds. When you cut off the first crop, side buds stem out from the leaf axil just below the cut.

STORAGE

Fresh broccoli can be stored in an airtight package 5 days in the refrigerator. Longer storage causes leaves to discolor and the stems to toughen. Canning is not recommended because of broccoli's strong flavor.

STORING LEFTOVER SEEDS

Under suitable conditions over 50% of the seed will germinate after 2-3 years of storage.

BRUSSELS SPROUTS

**Brussels Sprouts Or "Sprouts"
(No Other Common Names)**

***Brassica oleracea* Gemmifera Group**

Cruciferae (Mustard Family) A "Cole Crop"

Cool Season Biennial, Grown As An Annual

Killed By Prolonged Freezing Weather

Full Sun Or Partial Shade

Ease Of Growth Rating -- 4 to 6; Depending On The Length Of The Cool Growing Season

DESCRIPTION

Brussels sprouts are closely related to cabbage but, instead of forming a central head, make numerous buds and long-stemmed leaves on an erect stalk. The stalks average 16 to 30 inches or more in height, depending on the variety and the length of the cool growing season. The sprouts (buds) mature "from the bottom up" and the bottom ones are picked first. In long season areas it is not unusual to see a plant with a bare stem of 18-24 inches where the sprouts have been harvested.

Brussels sprouts prefer a long, cool growing season and, where summers are cool, are best suited for spring planting. The crop takes so long to mature that it is not ready for harvest until late summer or fall. In warmer areas of zone 3, and south through zone 7, planting from midsummer through fall is recommended because the plants will mature during cool, moist fall weather and will continue to produce well into the winter. Spring plantings in the latter areas rarely yield a worthwhile crop before being killed by hot weather.

History suggests that Brussels sprouts were selected from some form of cabbage, a more ancient vegetable. Sprouts have an advantage over cabbage in that only enough for a meal need be harvested at one time, whereas most cabbage heads will provide for several meals.

SITE

Brussels sprouts should not be grown in soil where other cole crops have been planted during the past three years. Spring planted sprouts need a location with full sun all day. Late summer planted seeds or transplants benefit from light shade in the juvenile stage, which reduces sunburn and wind damage, especially in zones 6 and 7. Tall, dense vegetables can cast a significant amount of shade from the afternoon sun; if possible, plant your seedlings on the east side of such a row.

SOIL

Brussels sprouts will tolerate a wide range of soils, but prefer sandy loam or well-drained clay loam soil with pH ranging from 5.5 to 6.8. Sprouts will tolerate somewhat alkaline soil, but the growth will not be as rapid. Good drainage and a quick drying soil surface work in favor of Brussels sprouts because, when fairly closely spaced, the tops of plants form a canopy which shades the ground and retards evaporation.

SOIL PREPARATION

Incorporate moderate to heavy amounts of organic matter and pre-plant nutrients when preparing the soil. Brussels sprouts are in the ground for such an extended period that thorough preparation is required in order to maintain moisture and nutrient holding capacity throughout the season. Dig or till-in generous amounts of organic amendments 2 to 3 weeks prior to planting so that they will have an opportunity to break down even further, or prepare the bed for spring sprouts the previous fall.

PLANTS PER PERSON/YIELD

Grow 4 to 5 plants for each person. This should give you a total crop of 6 to 8 lbs. of sprouts over the harvest season.

SEED GERMINATION

Brussels sprouts seeds are strong and sure-sprouting. The usual seed germination is 70 to 85 percent. Seeds will come up reliably in soil as cool as 50°F but the optimum temperature is 70-75°F.

Soil Temp.	50°F.	60°F.	65°F.
Days to Germ.	14 days	10 days	6 days
Soil Temp.	70°F.	75°F.	80°F.
Days to Germ.	5 days	4 days	3 days

It might appear that the higher range of temperatures is favorable. This is not the case because of reduced seedling vigor due to the heat.

DIRECT SEEDING

In zones 1, 2 and the northern half of 3, direct seeding will work, because the plants have all spring and summer to grow and set on a good crop of sprouts for fall harvest. Yet, many gardeners in these zones prefer starting with well-grown transplants in order to bring in a crop earlier in the fall. Zones 4 & 5, a crucible for cole crops, have too short a spring and too hot a summer for direct seeding in the spring and too short a fall season for a long harvest. Direct seeding in late summer or early fall in zones 6 and 7 is difficult because of hot, dry soil. Sow seeds ½ inch deep and 4 to 6 inches apart. In warm or hot soil, covering seeds with vermiculite after planting in the bottom of a water-saturated furrow will insulate them and increase their percentage of germination.

TRANSPLANTING AND THINNING

Set transplants or thin seedlings to stand 18 to 24 inches apart in the row and space rows 24 to 36 inches apart. Use the closer spacing where gardening is intensive and uniformly high levels of soil moisture and fertility are maintained.

Started plants of sprouts may be difficult to find because they rank behind other cole crops such as cabbage and broccoli in popularity. You may need to grow your own, and this is not difficult if you start early to produce large, strong plants for setting in the garden. The earlier you transplant, the more careful you must be with hardening off plants. Any kind of shelter to break the force of cold spring winds will help speed adjustment to the garden. As with broccoli, Brussels sprouts transplants should be set into garden soil a bit deeper than they were in the pot or flat, but not buried up to the first set of branches.

Summer transplanted Brussels sprouts need special care to reduce shock. Water and feed the seedlings before transplanting and dig holes in advance. Fill holes once or twice with water and let it soak in. Trim off the top 1/3 of the foliage to reduce transpiration loss. Set the plants in immediately after removing from the container. Pull soil up around them and water again to settle the soil around roots. Don't pack the soil down around the roots. Stick a shingle at an angle into the soil on the west side of the plant to give it some shade from midday and afternoon sun. Transplant in the evening or on a cloudy day, if possible. Water transplants daily for at least a week, not just a sprinkling, but an honest flooding.

MULCHING AND CULTIVATION

Summer planted sprouts can be mulched with organic litter to keep the soil cool. Spring plantings can be mulched with black plastic, but the improvement is marginal. Avoid compacting the soil around sprouts and cultivate plants grown under clean culture by scraping rather than hoeing to remove weeds. Deep hoeing can injure surface feeder roots.

SUPPORTING STRUCTURES

Tall growing varieties can become top heavy. In areas subject to driving rains or heavy winds, each plant should be securely tied to a rigid stake driven at least one ft. into the ground. Tie the plant above and below the concentration of buds to keep it from slumping.

WATERING

During dry periods, irrigate Brussels sprouts at least weekly, depending on the water holding capacity of the soil. The soil should be kept moist but not waterlogged. Furrow irrigation reduces the chance of diseases and the splashing of soil into the buds. Apply 1 to 2 inches to wet the soil to a depth of 5 to 10 inches. It is especially important to keep young transplants adequately watered while they are regenerating feeder roots.

CONTAINER GROWING

Brussels sprouts are superb plants for container growing. Individual plants will fit into 3 to 5 gallon containers. These may seem to be large for one plant, but you will soon see that the container is in proportion to the size of the mature plant. Frequent feeding and watering will be necessary.

Spring crops of Brussels sprouts are possible in containers in zones 1, 2 and part of 3, and along cool western coastal slopes, but over the balance of the country your best bet is to go with a crop planted in late summer for winter harvest.

ENVIRONMENTAL PROBLEMS

One rather subtle problem which shows up only when plants fail to mature as they should, is transplanting shock. If plants are started indoors and allowed to become rootbound or shocked by not protecting them properly during the adjustment period in the garden, maturity can be delayed by as much as two weeks, which can be critical in a summer planted crop.

HARVESTING

Snap or twist off sprouts at a diameter of 1 to 1½ inches. At this stage the buds should be firm, compact and bright green. Pinch buds gently before pulling; don't take the ones that are still fluffy.

To concentrate the maturing of sprouts, pinch off the growing point when the bottom sprouts are ½ to ¾ inch in diameter. This will terminate vertical growth and encourage the upper sprouts to develop faster.

Brussels sprouts can be harvested over an extended period during the winter in zones 1-4 if they are protected from destructive freezing and thawing by straw pulled up around the plants. An even longer harvest period can be enjoyed if the plants are dug with soil around the roots and stacked standing up under a shelter or in a cool root cellar. In zones 5-7 you should be able to harvest sprouts well into the winter but late January and February, hard freezes may wipe out the plants.

STORAGE

Brussels sprouts may be kept in the refrigerator in a moisture proof bag up to 5 days. Because of their strong flavor, brussels sprouts are not recommended for canning.

SEED STORAGE

Seeds placed in a sealed container with a desiccant and stored in a refrigerator should have a germination rate of at least 50 percent after two to three years of storage.

CABBAGE

Cabbage (No Common Names)

Brassica oleracea Capitata Group

Cruciferae (Mustard Family) A "Cole Crop"

Cool Season Biennial, Grown As An Annual

Killed By Prolonged Freezing Weather

Full Sun Or Partial Shade

Ease Of Growth Rating, When Spring Planted -- 4;
When Planted Later For Fall Harvest -- 3

DESCRIPTION

Everyone knows what grocery store cabbage looks like, but not everyone recognizes it growing in the garden. But anyone can distinguish the taste of freshly harvested cabbage from that which has been stored for some time.

Cabbage is easy to grow in all regions if planted to mature during relatively cool weather. Cabbage is grown for its heads which may be eaten as soon as the leaves begin to fold over into tight buds, or held in the garden to full maturity.

Certain spring-planted varieties of cabbage will split and become useless soon after heading up; others will hold for over a month with little loss in quality. A prolonged spell of above-average rainfall just after cabbage has headed can cause splitting. Bursting may also be associated with flower and seed formation in late winter, and during spring of the year following planting. The later, slower-growing varieties of cabbage rarely split.

New gardeners are invariably surprised by the spread of mature cabbage plants, which can range up to 3 feet in late, large-headed varieties. Midget varieties, on the other hand, may spread to cover only a 1 foot circle at maturity.

Four basic classes of cabbage varieties are available to home gardeners.

EARLY-MATURING SPRING CABBAGE, usually with small round or conical heads, light to medium green in color.

MIDSEASON TO LATE-MATURING CABBAGE with some varieties in the blue-green color range and others in medium-green shades. Varieties can be round, oval, flattened or drum shaped. Midseason to late varieties can mature as much as two months later than spring varieties.

SAVOY CABBAGE, which has light green to blue-green heads and waffle-textured leaves and makes marvelous coleslaw.

RED CABBAGE, a welcome change in cabbage color and flavor.

The more attention you pay to selecting varieties adapted to your area and planting them at the right time of the year, the easier it will be for you to grow cabbage. The two major planting seasons are early spring and midsummer through fall. Problems with splitting of heads, disease and insect pests, and mediocre quality have convinced most gardeners to avoid planting cabbage for midsummer maturity except in zones 1 and 2 where summers are cool.

Heading cabbage was not developed until the fourteenth century but, for hundreds of years prior, non-heading cabbage was grown by the ancient Britons. Wild cabbage can be found in the British Isles and northern Europe. Cultivated cabbage was apparently introduced across Europe during the Roman conquest.

SITE

Cabbage prefers full sun all day, except fall crops in zones 6 and 7, which benefit from light shade. Some garden books claim that cabbage will tolerate shade elsewhere, but it does so at the expense of quick, compact, growth.

Avoid planting cabbage where any cole crop has been planted during the past 2 or 3 years. Crop rotations lessen the risk of soil-borne diseases.

SOIL

Cabbage will tolerate only slight soil acidity, and grows well at a pH range of 6.0 to 6.8.

SOIL PREPARATION

Cabbage will grow well on a wide range of soils. Gardeners who have heavy clay soils should set aside a section of their garden for spring crops such as cabbage. In this section, build frames or raised beds and modify the top two or three inches of the soil with sand and/or well-decomposed compost to lighten the soil and make it easier to work. Sand is usually more expensive than home produced compost but warms up and drains faster, promoting faster growth early in the spring. Comparable results can be achieved by ridging up heavy soils and planting cabbage seedlings on top of the ridges. No such special culture is needed for sand or sandy loam soils.

Work in plenty of organic matter, but do not include compost containing trimmings from last year's cole crops. Such compost can transmit cabbage diseases from infected plant debris.

Spring cabbage needs more of the nitrate form of nitrogen than summer planted cabbage because spring cabbage is planted when the soil is cold and the natural release of nitrogen from the organic matter in the soil is proceeding slowly. Most "natural" or organic fertilizers contain nitrogen in the ammoniacal form. Organic gardeners can hasten the conversion of ammonia to nitrate forms of nitrogen for stimulating spring cabbage by incorporating natural nitrogen sources in compost and covering the pile with clear plastic to act as a solar trap. Within the warm compost, conversion from ammonia to nitrates will go on rapidly.

PLANTS PER PERSON/YIELD

For fresh use, grow 4 to 6 plants per person to yield 10 to 15 lbs., providing some of the heads are harvested when half grown. Harvest weights can vary considerably, depending on the variety and its stage of maturity at harvest time.

SEED GERMINATION

Cabbage seeds will sprout in cool soil and at high soil temperatures as well. However, the seedbed environment at these extremes can be hostile and the seedling mortality greater than at the optimum temperature of 70 to 75°F. Cabbage for fall harvest in zones 1-3 is often planted in soil as warm as 80°F with no decrease in seedling vigor but, further south, where late summer soil temperatures often exceed 90 to 100°F, planting or transplanting of cabbage is usually delayed until the heat abates.

Soil Temp.	59°F.	68°F.	77°F.
Days to Germ.	9 days	6 days	5 days

Soil Temp.	86°F.	95°F.
Days to Germ.	4 days	No Germ.

DIRECT SEEDING

Direct seeding of cabbage works best for later plantings, from midsummer in the North to early fall in zone 7. The first-early plantings in all zones except along the cool Pacific Coastal Plain are usually started from transplants.

When direct seeding, plant a pinch of two or three seeds in a group, covering them to a depth of ¼ inch. Make the groups 15 to 18 inches apart.

TRANSPLANTING AND THINNING

Spring-planted cabbage is almost always started from transplants because direct-seeded cabbage generally matures 2 to 3 weeks later. Plantings for fall harvest are also often started from transplants to shorten the planting-to-harvest span.

Some gardeners prefer to start their own plants indoors from seeds sown 5 to 8 weeks before the time to set them in the garden. They grow the seedlings in full sun or under fluorescent lights and thin or transplant to provide 2 to 3 inches between the seedlings in order for them to fill out properly.

The earlier you transplant to the garden, the greater the need for hardening. Set the seedlings out of doors in a protected area to accustom them to wind and cold. Even though well-hardened cabbage seedlings are frost hardy, they can be lost to severe freezing weather when there is no snow cover. Reserve part of the seedlings for replanting in that event.

When transplanting, set seedlings a bit deeper in the garden soil than they grew in the container, but not so deep that the bottom leaves are buried. Discard any seedlings that have "lodged" (fallen over) and regrown into a kinked plant.

Transplant seedlings 15 to 18 inches apart. This may be closer together than optimum for large, late varieties. If plants start to crowd each other, remove every other plant for early table use. Row spacing depends largely on the mature size of the variety and can run from 18 to 30 inches.

When transplanting, avoid compressing the soil around seedlings. Soil compaction can retard root growth.

MULCHING AND CULTIVATION

In zones 1 and 2 consider applying a clear plastic mulch over the cabbage bed a few weeks before transplanting spring seedlings. The clear plastic will trap more heat than black sheeting, and will shed the cold rains that delay soil warming. Cut slits for inserting transplants. Within this solar trap the warm soil will support a flourishing growth of the bacteria that convert the ammoniacal nitrogen of natural fertilizers into the nitrate form, available to plants. You will be pleasantly surprised at the difference made by mulching with plastic as compared to cabbage given clean culture which has to wait until the soil warms up for adequate nitrate nitrogen to become available.

Weeds will grow under the "greenhouse" made by the clear plastic; let them grow for a while then lay old cardboard or newspapers over the plastic to shade them out.

Black plastic is the customary mulch in zones 3 and 4, but is rarely used in zones 6 and 7 because the soil can get too hot beneath it at the time cabbage is heading.

In all zones, organic mulches are preferred for summer-planted cabbage, because they keep the soil cool and moist around the seedlings. If sawdust or wood shavings are used for mulching, first compost them with a little organic or synthetic nitrogen fertilizer to start the breakdown process.

Cabbage has a rather shallow root system, and weeds should be scraped or pulled from around them. Deep cultivation should be avoided as it can injure the roots. There is no need to pull up soil around cabbage unless, by error, you transplant seedlings too shallow and they tend to flop over.

CABBAGE

WATERING

Water cabbage transplants twice daily until they are established. If your soil is heavy and inclined to shed water, make little basins around the transplants to help the water soak in. Keep garden soil moist but not waterlogged. If no rain falls for 5 to 7 days, irrigate with 2 to 4 inches of water to wet the soil to a depth of 10 to 20 inches. In cool areas, apply only half as much water; evaporation and transpiration are reduced due to cooler soil and air temperatures. Furrow or drip irrigation is preferred; sprinkling can help spread disease organisms.

CONTAINER GROWING

Cabbage makes an excellent container plant and looks best when 3 to 4 heads are grown in containers of 5 gallon capacity. Even more heads, up to 5 of the small or midget varieties, can be grown in 7 gallon foot tubs. When the heads begin to crowd each other one or two should be removed to make room for the others to expand. If you plant some of the cabbages near the rim of the container they will hang over the edge, permitting you to cram more heads in each container. Since you are going to the work and expense of purchasing or preparing special soil mixes and filling containers, it is prudent to buy seeds of modern disease-resistant varieties rather than relying on run-of-the-mill varieties which may be all that are locally available as plants.

You can plant cabbage seeds directly in the container quite early and cover with a hood of clear plastic, which will serve as a solar trap, make the seeds sprout quickly and develop so rapidly that they will catch up with transplants.

ENVIRONMENTAL PROBLEMS

Cabbage is quite frost-hardy, but only if the plants are hardened off or acclimated to cold weather and brisk winds before being transplanted to the garden. Cabbage can also suffer from excessive hardening off, usually caused by commercial growers starting the plants too early and keeping them in small containers so long that the plants become hard, wiry and slow to resume growth. If your retail source has only this kind of abused plant, do not buy them; look elsewhere.

Transplanting cabbage plants too deep into the soil can delay maturity and can even cause them not to set heads. Go by the change of color on the stem indicating the soil line in the flat or pot. Don't bury the plant by covering the lower leaf stems.

Except in zones 1 and 2, the most prevalent problem with cabbage is planting it too late — so late that the heads will have just begun to wrap up when hot weather comes, with all its attendent problems. Hot weather stresses cabbage and makes it use so much water in the cooling process that it cannot reach or maintain good quality.

Cabbages can bolt in late fall and winter if subjected to extremely variable weather. Certain varieties are noteworthy for their resistance to splitting and bolting under such changeable weather conditions.

In zones 5 through 7, late cabbage is often planted in time to begin heading before cold weather sets in. During most years, the heads will overwinter and can be harvested as needed. "Slow bolting" varieties are recommended to avoid problems with heads splitting and shooting up seed stalks during periods of warm winter weather.

If you wait too late in the fall to plant cabbage in zones 5 and 7 it may not begin to head up before cold weather, and will go through the winter in the seedling stage. Such delayed crops are subject to bolting before full head size is reached in the spring.

HARVEST

When heads are firm, the cabbage is ready for harvest. Heads may be small, no more than 3 or 4 inches in diameter, or may be left until they reach full size. Don't wait until heads split or begin to lose their color. Use a long, sharp knife to cut the stalk just below the head. Early cabbage will sometimes resprout from buds on the stub and will produce a second crop of smaller heads. Your chances of having a good second crop of small heads are better if you leave several basal leaves on the stem and if you cut the stem on an angle so that the stub will shed water.

STORAGE

Cabbage may be stored in the refrigerator in a moisture-vapor proof bag for 1 to 2 weeks. Cabbage is a strong flavored vegetable and preserving is not recommended except as kraut for canning.

STORING LEFTOVER SEEDS

Cabbage seeds usually germinate 75-90 percent. Seeds placed in a sealed container with a desiccant and stored in a refrigerator should germinate 50 percent after three years.

CHINESE CABBAGE

Chinese Cabbage, Also Including Chinese White Cabbage, Chinese Flowering Cabbage And Chinese Flat Cabbage

Non-Heading Chinese Cabbage (Pak-Choi, Bok Choy, Tientsin Cabbage)

Brassica rapa Chinensis Group

Heading Chinese Cabbage (Celery Cabbage, Napa Cabbage, Pe-Tsai)

Brassica rapa Pekinsis Group

Both Are Cool Season Biennials, Grown As Annuals

Killed By Heavy Frosts

Chinese White Cabbage (Called Paak-Tesoi Or Pe-Tsai In China And (Confusingly) Bok Choy In San Francisco's Chinatown

Also Called Pak Choi-Bok Choy In Some Catalogs

Brassica chinensis Var. *chinensis*

Flowering White Cabbage (Called Paak Ts'soi Sum In China And Choy Sum In San Francisco's Chinatown)

Brassica chinensis Var. *parachinensis*

Chinese Flat Cabbage (Called Taai Koo Ts'oi In China And Tai Koo Choy In San Francisco's China-town; Referred To As Spoon Pak Choi Or Spoon Cabbage In Some Catalogs; Watch This Item: Chinese White Cabbage Seed Is Sometimes Used By Well-Meaning Seedsmen Who Don't Know The Difference)

Brassica chinensis Var. *rosularis*

All Three Are Cool Season Biennials, Grown As Annuals

Chinese White Cabbage And Flowering White Cabbage Are Killed By Heavy Frosts But Chinese Flat Cabbage Will Usually Survive Winters In Zones 4 through 7

Cruciferae (Mustard Family)

Full Sun Or Partial Shade

Ease Of Growth Rating, When Grown For Fall Or Winter Harvest -- 2; When Spring Planted -- 3

DESCRIPTION

This group of dual purpose salad greens/potherbs referred to loosely as the Chinese cabbages is complex and confusing. First and foremost, they neither look nor taste like ordinary cabbage.

Plant scientists, even Oriental taxonomists, disagree when classifying the various kinds by species and groups. It isn't much help to ask the Chinese or Chinese-American growers who produce the crops commercially because they use colloquial names, sometimes interchangeably between groups. Our attempt to create order from this chaos could be likened to trying to nail jelly to the wall. Only the lumps might hold, but not for long.

If you ask, "Why bother?", let us explain that throughout Asia the Chinese cabbages are a major food item cooked alone or with bits of pork, fowl or seafood. They are among the most space-efficient vegetables, and certain of the groups may be eaten either raw in salads or sandwiches as a lettuce substitute, or cooked. Each group has its own distinctive flavor and texture.

The heading types of Chinese cabbage can be stored in a cool, moist area for an extended period during winter.

The Chinese cabbages, now a minor vegetable, are destined for greater importance as Americans and Canadians continue to experiment with more vegetables in their diets. Here are descriptions of the five groups available in North America:

CHINESE CABBAGE, NON-HEADING

The medium green, somewhat rough surfaced leaves gather at the center to form a loose cluster. At maturity the center of this cluster blanches snow white or light golden, depending on the variety. The plants reach about 6 inches in height and 12 to 16 inches across. This type has leaves much like the heading kind but the plants never form compact heads. Bok Choy (the most often used common name) will tolerate more heat than the heading varieties but is more sensitive to cold. It does not store well in the garden nor in root cellars.

CHINESE CABBAGE, HEADING

Best known as "Celery Cabbage" or "Napa Cabbage" the popular heading group includes varieties with tall, slim, cylindrical or tapering heads as well as varieties with short, squatty, barrel-shaped heads. The tall cylindrical varieties generally have medium to dark green outer leaves while the round-headed varieties have lighter yellowish-green leaves fading to cream in the center of the head. The veined leaves are large and thin, tightly crumpled and packed into medium-firm heads.

Plants can reach 12 to 18 inches or more in height, depending on the variety and can grow to 24 inches across. Plants can be crowded within the row if sufficient space is allowed between rows for the lower leaves to spread out.

CHINESE WHITE CABBAGE

Grown almost exclusively as a potherb, this vegetable tastes somewhat like regular Chinese

cabbage but has a different texture, and superficially resembles Swiss Chard. The plants have upstanding, ladle-shaped white stems, clasping at the base. The inner leaves are not blanched. It is an elegant little vegetable that will grow 12 to 18 inches high and to an equal spread. Prefers cool weather.

FLOWERING WHITE CABBAGE

Grown more for its flowering shoots than for leaves, it is harvested in winter or spring just as the yellow flowers begin to open. A few tip leaves are also taken with the flowering shoot. The taste is mild; the texture tender, if chopped and stir fried to leave the vegetable a little crisp. The mature size of the plant depends greatly on the time of planting but would average 18 inches by 12 inches in spread. In China, this is an expensive delicacy because only the shoots, and not the entire plant, are harvested to be cooked or pickled.

CHINESE FLAT CABBAGE

Rarely seen in North America but popular in parts of China because of its winter hardiness. The plants are distinctly flat and low growing, with long, straight, pale-green leaf stalks tipped with dark green, glossy, round leaves the size and shape of a tablespoon. At maturity the plants are only 2 to 3 inches tall but spread to 14 to 16 inches across. When grown for fall and winter use in zones 4 through 7, the plants should survive the winter. Prefers cool weather.

SITE

Spring planted Chinese cabbage needs a sunny location away from frost pockets that set back growth and induce premature flowering. In zones 5 through 7, late summer or fall plantings of Chinese cabbage appreciate afternoon shade if it can be supplied without competition from tree roots.

SOIL

Moderately fast draining soil, rich in nitrogen and well fortified with organic matter, will produce fast growth. Chinese cabbage will thrive in a wide range of soil conditions, from pH 6.0 to 7.6.

SOIL PREPARATION

Some gardeners prefer to roughly prepare beds for Chinese cabbage in the fall, leaving them covered with manure or coarse compost. This is turned under in the early spring and the beds leveled for seeding. In areas where spring rainfall is heavy, the soil is generally ridged up or formed into raised beds to permit water to drain away in the pathways. If your raised beds are three to four feet wide, dish the centers to simplify watering.

Cabbage family members can contract a disease called "Clubroot", characterized by knots on the roots, not to be confused with nematode damage. A typical recommendation for avoiding the situation is to raise the soil pH to at least 7.2 by applying hydrated lime about a week before planting. It works, but it can cause unexpected nutrient imbalances, most temporary. Precision liming is not easy and dealing with the side effects is even more difficult . . . yet it is preferred to either not growing cole crops and Chinese cabbage, or suffering the poor performance that comes from untreated soil. The best solution is to work closely with your County Agricultural Agent (Cooperative Extension Service) to develop an integrated liming/nutrient control program.

Clubroot is not a problem on high pH western soils.

Do not bare-root transplant this crop. It is very difficult even for experienced growers. Peat pots or Seedling flats or some other method where roots are not disturbed at all are suggested. All of these plants will bolt when they are transplanted as bare-rooted seedlings.

PLANTS PER PERSON/YIELD

Chinese cabbage matures all at once. Begin eating it when the plants are only half grown to keep your crop from being a transient pleasure. In spring plantings, heads will hold for only a week or two before shooting up seeds, so one or two mature heads per person would be sufficient. Fall Chinese cabbage will hold in good condition for weeks and you could grow about six heads per person, more if you plan to store the excess in a root cellar. Heads average about 3 pounds each.

You can grow more plants of Chinese white cabbage, flowering white cabbage and Chinese flat cabbage than of the regular Chinese cabbage because these three can be harvested leaf by leaf; you need not take the entire plant.

SEED GERMINATION

The optimum soil temperature for germinating seeds in this group is 60 to 65°F., but seeds will sprout within the range of 45 to 75°F. The usual germination rate is 70 to 80 percent.

MULCHING AND CULTIVATION

The growing of all of the Chinese cabbages must be geared to their habit of flowering as the days grow longer. By its nature, Chinese cabbage is suited for heading as the day length lessens, toward fall, although a few recently introduced varieties are available which will head as the days increase in length and in the long days of early summer in the North. Although Chinese cabbage prefers to grow in cool weather, temperature has much less affect on the bolting of plants than the length of nights.

Spring-planted chinese cabbage responds well to plastic mulches. In zones 1-3, mulches of clear or black plastic over built-up beds can raise the soil temperature several degrees and activate the bacteria that convert ammoniacal organic nitrogen to available nitrate forms.

Use coarse organic materials such as straw, hay or dried grass clippings on top of sheets of newsprint or cardboard to mulch summer-planted crops.

Chinese cabbage is rather simple to cultivate because the plants grow rapidly and are easily distinguished from weeds. Hoeing should be shallow; scraping away of weeds is preferred to chopping.

WATERING

Chinese cabbage is a hog for water, especially during windy, dry weather. While it prefers well-drained soils, it also needs a uniformly high and consistent level of moisture. This situation naturally exists in soils containing a good content of humus which can act as a reservoir for moisture while keeping the soil open for good drainage. If no rain falls for 4 to 5 days, irrigate with to 1 to 2 inches to moisten the soil to a depth of 5 to 10 inches. Apply twice this much if your water is cheap; it won't be wasted.

CONTAINER GROWING

Chinese cabbage makes an excellent container plant. One plant of the heading or non-heading regular Chinese cabbages will fill a 3-gallon bucket. The heading types are too large to plant around the base of summer vegetables in containers, but you can do this with the smaller kinds such as Chinese white cabbage if you will harvest the plants while they are still young.

ENVIRONMENTAL PROBLEMS

Premature flowering can be a serious problem where seedlings are exposed to long periods of freezing or just-above-freezing weather. For this reason, planting is usually delayed until mid-spring in zones 1 through 4.

HARVESTING

Heading types are ready for harvest when the firm, cylindrical or oval head has developed. Then the entire plant is pulled. The roots and outer or wrapper leaves are discarded. They are usually weatherbeaten and tough.

Non-heading types of Chinese cabbage, Chinese white cabbage and Chinese flat cabbage are grown for their succulent leaves and stalks which may be cut one at a time, or the entire plant may be harvested if desired.

Flowering white cabbage, which is grown for its shoots, will stay in production for only 2 to 3 weeks in late spring before the plants deteriorate. Harvest two or three tip leaves along with the flowering shoot and leave the plant to produce more shoots. A reminder: Flowering white cabbage should always be winter or spring planted; it will not produce flowering shoots if planted in late summer or early fall because lengthening days are required to produce flowering shoots.

STORING LEFTOVER SEEDS

The seeds of all kinds of Chinese cabbage are relatively long-lived. When placed in a sealed container with a desiccant and stored in a refrigerator, seeds should germinate at least 50% after three years.

STORAGE

Chinese cabbage may be kept in the refrigerator in a moisture proof container up to 2 weeks. Heading types of Chinese cabbage can be stored in the garden for several weeks if grown for fall harvest, because in the shortening days of fall they will not go to seed. When heavy frosts are threatening, the plants can be covered with leaves or straw to protect them temporarily or with a canopy of clear plastic over wire hoops to add several days to the harvest period.

Heading types of Chinese cabbage, the barrel-shaped "Bok Choy" varieties, will store for 4 to 6 weeks in a cool, moist root cellar; the tall cylindrical types even longer. Dig the plants, roots and all, and set them upright on a bed of moist sand. Do not sprinkle the tops of the heads as this could cause them to rot.

CARROTS

Carrot

(No Common Names)

Daucus carota **Var.** *sativus*

Umbelliferae (Parsley Family)

Biennial, Grown As An Annual

Cool Season

Killed By Prolonged Freezing Weather

Full Sun Or Partial Shade

Ease Of Growth Rating -- 4

DESCRIPTION

Carrots can be grown in all North American climate zones. They have feathery tops that grow 10-16 inches in height and up to 18 inches across, and extensive root systems, with deep anchor roots and shallow feeder roots. Usually only the roots of carrots are eaten, and since they are eaten raw as often as cooked, great attention has been paid by plant breeders to improving the taste, tenderness, and appearance of home garden varieties.

Until the beginning of the twentieth century carrots were used primarily for medicinal purposes or livestock feeding and not generally as a vegetable. Their origin can be traced to carrots with purple roots which were domesticated in Afghanistan and which spread to the eastern Mediterranean area and into western Europe in the fourteenth and fifteenth centuries. By the sixteenth century, roots in a wide range of sizes, colors, and shapes had been developed. Until early in the nineteenth century, carrots in North America were known more for their quantity than their quality, and the very large, coarse roots were harvested for stock feed. Europeans have long been more attracted to carrots as a delicacy than Americans and cultivate many varieties, each with its loyal constituency.

SITE

Carrots should have full sun all day. Except in the deep South and desert areas of the West, where summer sunshine is intense. There, carrots should be given light afternoon shade during midsummer. Avoid competition from roots of trees and shrubs.

SOIL

Carrots will grow well at soil pH levels of 5.5-6.8. Commercial growers of carrots look for sandy soil or organic muck soil for producing good crops of long rooted varieties. Such soils have several advantages: carrot seeds will emerge readily since these soils do not crust; the roots will grow long, straight and smooth; and they can be pulled from lightweight soils with little resistance. You may not be able to duplicate these conditions in your own carrot bed without several years of soil improvement, but you can come close to it the first year by thoroughly incorporating 3 inches of well decomposed organic matter to spade depth. In heavy clay soils sand can also be added.

SOIL PREPARATION

A deep seedbed is needed to grow long, large carrot roots. Turn under and incorporate well-decomposed, screened organic matter, pre-moistened peatmoss, or pasteurized packaged manure to a depth of at least 12 inches. Break up the clods and remove stones and debris. In zones 5, 6, and 7, where organic matter rapidly oxidizes and disappears from the soil due to prolonged high soil temperatures, incorporate organic matter every year for a root crop such as carrots. Carrots can be planted as early in the spring as soil can be worked; therefore, it is good practice to prepare your carrot bed the previous fall. Dig in a combination of green and dry organic matter to a depth of 12 inches in late fall, along with manure and pre-plant fertilizer. Early in the spring, turn it over with a spading fork to a depth of 12 inches. By spring most of the organic matter will have decomposed considerably. Remove any large particles.

Building up beds is not necessary in naturally well-drained soils but, if your soil is heavy or poorly drained, build up carrot beds at least 4-6 inches above the level of the surrounding soil to improve the drainage and aeration. Make the beds about 3 feet across and plant 2 rows of carrots near the shoulders. Avoid walking on the bed between the two rows; soil compaction is injurious to carrots. On a 3 foot wide bed, if the soil is light and deeply prepared, you can plant a third row of carrots down the middle and harvest it early, leaving room for the 2 outside rows to develop.

PLANTS PER PERSON/YIELD

Grow 50-75 plants per person for fresh use. This should yield 6-8 pounds of carrots, if some of the roots are pulled at small sizes. Grow triple this amount if carrots are to be processed and/or stored for winter use.

SEED GERMINATION

Soil Temp.	41°F.	50°F.	59°F.	68°F.
Days to Germ.	51 days	17 days	10 days	7 days
Soil Temp.	77°F.	86°F.	95°F.	104°F.
Days to Germ.	6 days	6 days	9 days	No Germ.

The optimum soil temperature for germination is 80°F. The usual germination for carrot seed is 60-75%.

DIRECT SEEDING

Seedlings grow slowly and are susceptible to competition from weeds. As a result, seeds are often sown generously. The gardener may find an overabundance of seedlings if everything goes right, and will have to thin 2 or 3 times, often pulling out clumps of seedlings to leave room for the remainder to grow. Where carrots are not adequately thinned, virtually all the roots will be contorted. In heavy clay or dry sandy soils the opposite problem of too few seedlings can occur. Clay soils can seal over after rains making if difficult for seedlings to emerge. While dry soils can't provide the consistent level of moisture needed to soften the seed coat and energize the seedling within. Improved garden soils, such as those prepared by organic gardeners, have excellent moisture retention and will rarely crust over seeds. On most soils, you should cover the seeds with a non-crusting material such as sifted compost, vermiculite, or sand. (If your soil is sticky clay use sand.)

To plant the seed make a shallow furrow with the corner of a hoe and plant about 4 seeds per inch. Cover to a depth of ¼ inch — ½ inch if your soil is light and dry. Space single rows 12 to 18 inches apart. Carrots are perfect for wide row planting. Prepare a seedbed the width of a garden rake and scatter the seeds about 1 inch apart in all directions. Cover as previously directed.

In warm climate zones, when germinating seeds in late summer, water deeply before planting and cover seeds to a depth of ½ inch with porous material. It is important to keep the soil uniformly moist until the seedlings emerge, so cover the seedbed with burlap, cardboard or weighted layers of newspaper to reduce evaporation. Remove when seedlings appear.

TRANSPLANTING AND THINNING

Carrots are strongly tap-rooted and are rarely transplanted. Gardeners who abhor the waste in thinning often attempt transplanting when seedlings are small. Those who work fast with a gentle touch sometimes succeed but transplanting can set back maturity as much as 2-3 weeks.

As described under DIRECT SEEDING, thinning of carrots may seem wasteful, but it is necessary to produce large straight roots. Thin before seedlings reach 2 inches in height, the smaller the better. If you have a thin stand with only occasional thick clumps, water before thinning and remove the outer seedlings in each clump by pulling up and out to minimize damage to the remaining seedlings. It may be necessary to thin 2 or 3 times since carrot seeds tend to germinate over an extended period. Carrots should stand 1 to 2 inches apart on loose soil and 3 to 4 inches apart on hard clay soil. Those planted in wide rows should be thinned to stand 3 inches apart.

MULCHING AND CULTIVATION

Shallow mulches of organic matter can be hilled up to cover the exposed shoulders of the roots so they will not turn green, but mulching carrots is not recommended in parts of zone 6 and 7 because pests such as sowbugs, earwigs, slugs, and snails, tend to build up under them. Plastic mulch is not recommended for carrots because the roots grow so close together that you would have to cut a long slit in the plastic, which would weaken it and encourage tearing. Also, plastic mulch tends to invite walking over the root system of the carrots, which compacts the soil, reducing production and quality.

Hand weed frequently early in the life of your carrot crop. If the weeds grow large they will dislodge carrot plants when you try to pull them. Shallow cultivation is called for. Pull soil up from the row middlings as you cultivate, but no more than an inch at a time. If you plant in single or double rows you can hill up the soil over the exposed shoulders of the carrots to keep them from turning green.

Carrots' requirements are low for nitrogen, moderate for phosphorus, and high for potash. Phosphorus moves very slowly down through the soil and, in alkaline soils, can quickly be immobilized in the surface layer, where it is of little use to a deep rooted crop. Therefore, it is important to dig or till the major provision of phosphate deeply into the soil at the time it is prepared, using either organic or processed phosphate sources.

WATER

A uniformly good level of soil moisture is especially important during the seedling stage of carrots. Avoid letting carrots go for more than a week without rain or irrigation water. When irrigating, apply 2 to 4 inches of water to wet the soil to a depth of 10 to 20 inches. Furrow irrigation, by ponding water between rows, is easy and effective.

CONTAINER GROWING

Carrots make a great container crop. A 7 gallon tub is an appropriate container and will produce 10-15 carrot roots. Use a short rooted variety and thin when the roots have reached about 1 inch in diameter. Topdress the remaining carrots with growing media to keep the shoulders from turning green. Potash and nitrogen both leach out from fast-draining container mixes, but organic gardeners can compensate by top dressing with compost fortified with nitrogen and potash from natural sources.

ENVIRONMENTAL PROBLEMS

Beginning gardeners often have trouble getting carrot seeds to come up quickly and uniformly. Carrot seeds may require 2-3 weeks to germinate because of uneven soil moisture, but you can hasten

this by covering them with sand or very finely pulverized compost to a depth of ¼ to ½ inch, the seedlings can easily break through this porous cover.

If carrots are permitted to go dry for some time (and they can survive fairly long periods of drought), a certain percentage of roots is likely to split when they are given a heavy watering. To avoid splitting, do not allow carrots to go for more than a week without rain or irrigation.

Poor, winter color of carrot roots in zone 6 and 7 is due to low temperatures. Conversely, the failure of roots to reach the proper length in good soil may be due to high soil temperatures. Roots can fork or grow rough when raw manure or large chunks or undecomposed compost are present in the soil. Granular pasteurized manure is recommended to avoid this. Green or sun-roughened shoulders are a problem with certain varieties such as Nantes and Amstel. It may also occur where soil erodes away from the shoulders of roots which protrude from the soil. This is more than a cosmetic problem since the green color extends to the core and can give the top 1 to 3 inches of the carrot an "off" flavor.

To prevent the problem, pull the soil up deep enough to cover the tops of roots, but don't dig so deeply that you disturb the shallow root system. Stand to the side when hilling up carrots so that you do not compact the root system.

A frequent problem is that carrot roots may crack at or near maturity. One cause is the roots are being allowed to grow until they can't continue to do so without splitting, another is wide variations in soil moisture.

Snapping off of tops and leaving roots in the ground, is a common problem with beginning carrot growers (unless they have sandy or porous soil.) A spading fork or trowel should be used to loosen the soil so that roots can be pulled out without breaking.

HARVEST
Pull away some soil and check the root diameter. When the roots are ½ inch across, you can begin harvesting baby carrots. At this stage, the roots of standard varieties may be only 3 to 4 inches long. Pulling them will accomplish additional thinning, to the benefit of the remaining roots. The main crop of medium-sized carrot varieties will be ready when carrots are ¾ to 1 inch across the shoulder. Some carrot varieties become tough if harvested when the diameter exceeds 1 to 1½ inches. Pull carrots gently from moist porous soil. In heavier soils use a spading fork to loosen the soil before pulling.

Carrots sown for fall harvest may be left in the ground until needed, even until midwinter. In cold climate areas, carrots can be covered with a mulch for storage in the garden. The deeper the mulch, the longer the storage. If you place a sheet of plywood between layers of mulch you can tilt it up and remove carrots even when the top layers of mulch are caked with ice and snow.

STORAGE
Freshly dug carrots can be refrigerated in a moisture proof bag for several months.

STORING LEFTOVER SEEDS
When stored as recommended, carrot seeds should have a germination rate of at least 50% after 2 to 3 years.

CAULIFLOWER

Cauliflower

Brassica oleracea **Botrytris Group**

Cruciferae (Mustard Family), A "Cole Crop"

Cool Season Biennial Grown As An Annual

Killed By Medium Frosts

Full Sun Or Light Shade

Ease Of Growth Rating, From Transplants -- 7; From Seedlings You Grow From Seeds -- 8

DESCRIPTION

Cauliflower is closely related to cabbage, but rather than eating the leaves you eat the flower buds. Taken as a whole, the clustered flower buds are called the "curd" of the cauliflower. The ideal curd is tight, dense, smoothly pebbled, and free of what commercial growers call "rice", a picturesque term which needs no definition.

Cauliflower has not increased in popularity as much as broccoli because, when cauliflower is harvested, it will rarely "repeat" by producing lateral shoots. Being of a longer season of maturity than broccoli, cauliflower is more likely to be successful when planted for fall harvest. Cauliflower is more difficult to grow than cabbage or broccoli because of its decided preference for cool weather, and because of its low tolerance for checking of growth anytime during the juvenile period. Cauliflower seedlings are somewhat less frost tolerant than cabbage.

Cauliflower is such a close relative of broccoli that it bears the identical species and group name. Compared to its ancient relative, cabbage, cauliflower is probably of rather recent origin.

SITE

Grow cauliflower in full sun and in an area where no cole crops were planted the previous year. Spring crops of cauliflower benefit from planting on raised sandy beds, which permit the cauliflower to grow faster and to mature prior to hot weather.

SOIL

Well drained, sandy loam to clay loam soil is preferred for cauliflower. The pH should be in the range of 6.0 to 6.8, but it will grow on more alkaline soils if nutrients and micro-nutrients are in balance.

In extremely sandy soils it is often difficult to maintain a consistent level of moisture and nutrients. If you have such a soil, amend it highly with organic matter prior to planting cauliflower.

Cauliflower is especially sensitive to deficiencies of micronutrients (trace elements) which can show as hollow stems, discolored interiors or malformed curds which keep poorly.

SOIL PREPARATION

Avoid using compost which includes trimmings from last year's crops of the cabbage family. This could prove to be a source for plant diseases. Incorporate well-rotted compost, peat moss or purchased organic soil amendments to a depth of 8 to 12 inches, and give the soil a liberal application of mineral fertilizer or manure tea. If you plant cauliflower for fall harvest, following a spring crop of vegetables, the soil will likely be impoverished. Be sure to incorporate a source of nitrogen, because this is the most important nutrient element for forcing cauliflower to mature quickly. At all stages, provide cauliflower with growing conditions which encourage rapid, steady growth.

PLANTS PER PERSON/YIELD

Grow 4 to 6 plants for each person; this should yield 9 to 10 lbs. of cauliflower, but probably less because you will begin harvesting some heads at less than full size.

SEED GERMINATION RATE

Soil Temp.	50°F.	59°F.	68°F.
Days to Germ.	19 days	10 days	6 days

Soil Temp.	77°F.	86°F.	95°F.
Days to Germ.	5 days	5 days	No Germ.

Cauliflower seed is most often started indoors at a soil temperature of 65-75°F. It is important to drop the temperature to around 60-65°F. as soon as germination has occurred or the seedlings will stretch and become lanky.

DIRECT SEEDING

Transplanting is the preferred method for starting spring cauliflower, but many gardeners prefer to direct-seed the plantings they make for fall crops. Sow seeds in the garden 3 to 4 weeks prior to the summer transplanting dates indicated. Plant 3 to 4 seeds in a group, covering the seeds ¼ inch deep. Space the groups 18 to 20 inches apart. By planting more than one seed per location you are assured of at least one strong seedling and should not have to transplant to fill in gaps.

TRANSPLANTING AND THINNING

It is difficult to purchase cauliflower plants which have been grown without checking. Very often cauliflower seedlings are started too early and are hard and wiry by the time you buy them. These excessively hardened seedlings will sometimes form little "buttons" rather than full-sized heads. Consequently, many gardeners prefer to start their own cauliflower transplants from seeds sown indoors 4 to 5 weeks before the time to set them out into the garden. Start seeds at temperatures of 65-75°F. Keep the seedlings growing actively and transplant them into the garden before growth is stunted or checked. The hardening period should be rather short, no more than about one week. At no time should cauliflower seedlings be allowed to become potbound or allowed to go dry.

Thin cauliflower seedlings or transplants to stand 18 to 20 inches apart in rows 24 to 30 inches apart. The plants of cauliflower grown for fall harvest tend to be somewhat larger than spring cauliflower.

MULCHING AND CULTIVATION

Early mulching is not recommended for spring crops of cauliflower because it tends to keep the soil too cool and wet. However, black plastic can be applied in midseason for spring crops of cauliflower because it traps heat and keeps the soil around roots warm, hastening growth.

Mulching of fall crops of cauliflower with organic mulch or plastic sheeting is recommended because it will maintain a uniform level of soil moisture. In zones 5 through 7 the use of an organic mulch is preferred.

Weeding of cauliflower is usually done by hoeing. Be sure not to pull soil up around the plants, because covering the crown with soil can cause problems with the formation of heads.

All varieties of cauliflower except those with the "self blanching" character need the curd protected to develop a pure white, unblemished head. The procedure is called "blanching" and is done by gathering 3 or 4 of the outer leaves and securing the tips loosely with string or a rubber band. Tie when the curd is only the size of a large marble. Inspect the developing heads occasionally and treat with a safe for humans biological insecticide if cabbage worms are present. If blanching is done during a prolonged wet spell, let the heads dry occasionally to prevent rotting from trapped moisture.

WATERING

Cauliflower plants are sensitive to growth checks. They need to be kept growing rapidly and evenly. If no rain falls for 5 to 7 days, apply about 2 inches of water in order to wet the soil to a depth of 10 inches. You will rarely experience a growing season when you will not need to irrigate cauliflower in order to keep it growing rapidly.

CONTAINER GROWING

Cauliflower makes an attractive, interesting container plant. Single plants will grow in 3-gallon capacity containers but a good arrangement is two plants per 5-gallon container or three per 7-gallon tub. Fertilize the container soil every 2 to 3 weeks with a soluble organic plant food such as fish emulsion or with a dilute solution of high-analysis chemical fertilizer. If nutrient deficiencies are going to show up in any container-grown vegetable, it will be cauliflower. The major nutrients have to be in balance, and the secondary and micro-nutrients (especially boron) present in good supply. Lime needs to be added to most artificial soil media at the rate of ¼ cup per cubic foot of growing medium — mix thoroughly prior to planting.

ENVIRONMENTAL PROBLEMS

The most common problem with cauliflower is starting too late. Cauliflower requires about two months of cool weather to mature and, very often, gardeners are timid about setting out plants in very early spring when the weather is still quite cold. If you will acclimatize your cauliflower plants and protect them after transplanting with plastic jugs from which the bottoms have been removed, you can plant early and get a jump of three weeks on the season. You will rarely experience a crop failure when you expedite your spring cauliflower crop in this fashion.

Another common problem with cauliflower is using the wrong variety for the season or for your zone. Some varieties require longer than 60 days to mature, and this difference can be critical. Your cauliflower plants can be large and vigorous and the curd just beginning to form when hot weather comes along; this is the signal for cauliflower to go to seed and, regardless of what you do, the quality will degrade rapidly, curds will burst into flower and you will have nothing to eat for all your work.

HARVESTING

The edible head, or curd, is made up of numerous immature flower buds. Harvest by cutting off the entire head below the top set of leaves while the buds are still white, compact and firm. It is better to harvest cauliflower a bit early than too late, because it quickly becomes dried out and grainy.

Fall crops of cauliflower should be harvested if temperatures of less than 29°F. are forecast. If you can't freeze your harvest immediately, dig the plants, roots and all. Knock off surplus soil. Hang the plants upside down in a cellar; if the air is dry, spray the plants occasionally to keep them from drying out. This should maintain heads in good condition for about two weeks.

STORAGE

Cauliflower may be stored in the refrigerator in a moisture-vapor proof bag for up to one week. Because cauliflower is a strong flavored vegetable, canning is not recommended.

STORING LEFTOVER SEEDS

When stored in a sealed container with a desiccant and placed in a refrigerator, cauliflower seeds should germinate at least 50% after 2 to 3 years.

CAULIFLOWER

Celery And Celeriac

Celery (No Common Names)

Apium graveolens, Var. *dulce*

Celeriac (German Celery, Knob Celery, Root Celery, Turnip-Rooted Celery)

Apium graveolens, Var. *rapaceum*

Umbelliferae (Parsley Family)

Both Are Cool Season Biennials, Grown As Annuals

Killed By Prolonged Freezing Weather

Full Sun Or Partial Shade

Ease Of Growth Rating, Celery -- 9; Celeriac -- 8

DESCRIPTION

Celery is grown for its closely-clasped central stems, referred to collectively as a "stalk". The individual leaf blades are attached to the base by 5-10 inches long petioles called "stems." The foliage can be either green or golden yellow, with the latter especially evident in the inner leaves. Celery plants grow 14-24 inches or more in height and spread. The stalks you see in grocery stores are only the hearts of large, mounded plants. The many outer stems and abundant leaves are tough and stringy and though, not ordinarily used, are edible. Celery is a difficult vegetable to grow. Many gardeners have tried it and given up, not understanding its requirements for fertile, well-drained soil and a reasonably long, cool growing season. Prolonged cold periods can trigger bolting, a particular problem with celery.

Celeriac looks like small-stemmed celery, but forms a large, round, knobby root which partly protrudes from the soil. Usually only the root is eaten. Celeriac was selected from stalk celery's ancestors back when root vegetables were crucial to man's survival through the winter months. It is less demanding plant than celery, but subject to the same problem of bolting.

Wild relatives of celery and celeriac may be found from Sweden south to Ethiopia and west to India. Celery has escaped from gardens and grows wild in parts of California and New Zealand. Its first mention as a cultivated plant was in 1623 in France. Initially it was used as a seasoning, and it was not until the 18th century that mild-flavored types were grown for eating.

SOIL

Celery and celeriac like sandy to fast-draining clay loam soil or peaty muck soil, and a pH range of 6.0-6.8. It is advisable to maintain a consistently high level of soil fertility with frequent applications of liquid or granular fertilizer or manure tea. Western desert soils should be leached with 6-12 inches of water every 3 or 4 years to clean accumulated salts out of the root zone.

SOIL PREPARATION

Much of the "good-luck" experienced growers have in producing celery and celeriac can be attributed to careful and thorough soil preparation. Except on organic soils, with which few home gardeners are blessed, heavy amounts of organic matter should be incorporated to serve as a reservoir of moisture and plant nutrients. The organic matter can be mixed with the top 12 inches of soil. If you "double-dig" to a depth of 18-24 inches, put the excavated subsoil and topsoil in separate layers. Mix organic matter with each and return the amended subsoil to the trench first. The organic matter can be only half-rotted and unscreened, but be sure to remove large chunks which might impede the movement of water through the soil.

Incorporating heavy amounts of organic matter will add considerably to the volume of the soil and result in raised beds, a "plus factor" for celery and celeriac growing.

Celery and celeriac have different nutrient requirements but this can be compensated for by changing supplementary fertilizer rates . . . soil preparation can be the same.

In areas of high rainfall and heavy, slow-draining soils, plant celery or celeriac on raised beds approximately 4 feet wide. Mix in a layer of 2-3 inches of well-rotted manure or compost and, if the soil is known to be inherently poor, work in pre-plant fertilizer, either organic or manufactured. Plant 2 rows on the shoulders of the bed, leaving the center of the bed open for a wide irrigation furrow.

Another method is to build a frame to raise the soil 8-12 inches above the surrounding garden. Fill the frame with a mixture of equal parts of garden soil, coarse sand, and well-decomposed organic matter. Celery and celeriac will grow rapidly in this ideal situation, and clear slitted plastic can be tacked over the frame to insulate early-planted seedlings from cold snaps that cause bolting.

PLANTS PER PERSON/YIELD

Grow 10-12 celery plants for each person. Begin harvesting them when they are only half-grown, in order to avoid an excess of celery at the end of the season. A harvest of 12-14 lbs. of stalks is possible if you let all the heads mature. Grow 10-12 celeriac plants for each person, to get 8 to 12 lbs. of roots for fresh use and an excess for storage.

SEED GERMINATION

At 68°F celery and celeriac seed will germinate in 7 days. At 41°F it will germinate in 41 days. However, seeds in a given batch will sprout at different times. For this reason, commercial growers often soak celery and celeriac seeds in water in a warm (65-70°F) lighted room until the seeds sprout. Night temperatures are kept above 55°F to avoid bolting. As soon as the first sprouts appear, the seeds are spread on paper towels to dry enough for handling, and then immediately planted. To insure the effectiveness of presprouting, observe the following precautions. First, using a tea strainer, wash the seeds in a dilute solution of dishwashing detergent, and rinse them thoroughly. The detergent will remove or deactivate most disease organisms on the outside of the seed coat. Also, check the soaked seeds daily for signs of sprouting and move them to the seed flat as soon as the first sprouts appear. Do not allow the sprouts to grow long and stringy, as they will recover slowly (if at all) when covered with growing medium in the seed flat.

SITE

Reserve the best soil in your garden for celery or celeriac. Raised beds are recommended. In zones 1, 2, 3, and 4 grow celery and celeriac in full sun. In zones 6 and 7, where they can be planted in late summer for fall or winter harvest, they are often grown on the east side of taller vegetables such as okra or pole beans to take advantage of afternoon shade.

TRANSPLANTING

Large, well-grown celery or celeriac seedlings that have not been checked in growth are hard to find unless you buy from a greenhouse grower. Even then, there is a risk in very early spring that they have been exposed to several days of cold weather on a retailers display rack. They may have been accidently set to bolt before you have purchased them. It is advisable then to buy seedlings from lots delivered later in the season.

Some home gardeners prefer to start their own seedlings despite the long lead time and care required to do so. Start seeds indoors 10-12 weeks prior to the transplant date for your zone. Celery and celeriac seeds are small, slow, and uneven sprouting, and they require light for germination. Space about 6 seeds per square inch and cover them no more than ⅛ inch deep so light can penetrate to the seeds. It is, however, necessary to cover the seeds. If they germinate on top of the soil they can fail to orient themselves quickly and will not push down a root as they should.

Another method of starting seeds for transplanting is to pre-sprout (See SEED GERMINATION) and plant them in seed flats to reduce the chance of damping-off disease. Plant the pre-sprouted seeds when the first sprouts begin to show. Don't wait for the majority of the seeds to sprout. Whichever method you use, grow the seeds under fluorescent lights with the seed flat only 2 inches from the tubes. Keep the temperature near 70°F day and 60°F night, but never less than 55°F. Frequent watering with warm water will speed germination, which occurs in 2-3 weeks on the average. Pre-sprouted seeds will emerge much faster, of course.

When seedlings have developed 2 leaves, lower the temperature to 60-65°F and raise the light to 4-6 inches above the plants. While the seedlings are still quite small, pry them out with a pencil point and set them into individual 2 inch peat pots filled with a good grade of potting soil. Don't use large pots for tiny seedlings, for the soil tends to run too wet for them. The seedlings will be ready to transplant when they reach 4-5 inches high and the diameter of a pencil. Since they have been pushed to grow fast, using mild fertilizer solutions and lots of water in a warm environment, they cannot be transferred directly from such pampering to the cold, dry, and windy elements of your garden. They will suffer growth checking which delays their maturity. Find a warm, protected corner out of doors and keep the flats there for 7-10 days before transplanting. Reduce the frequency of watering.

To transplant, dig generous holes first. Fill them with water or a mild solution of liquid starter fertilizer and let it soak in. Set in the transplants to a depth matching the soil line in their container. Set celery transplants 8-10 inches apart in rows spaced 24-30 inches apart (or wider if placed on the shoulders of raised beds as described in SOIL PREPARATION). Space celeriac transplants 6-8 inches apart in rows 12-15 inches apart. Pull dry soil up around the seedlings and firm it down gently. To settle the plants, water again using a can of luke warm water brought from the house. Cold water from a hose can shock seedlings. Shelter the plants for a few days.

MULCHING AND CULTIVATION

Celery and celeriac respond well to mulching with organic or black plastic mulches. If you use the 4 feet wide raised beds, build the bed with the center dished before covering with sheet plastic. Cut holes for the celery and celeriac plants and one or two holes in the center to admit water. In zones 4, 5, 6, and 7 use organic mulches to keep the soil cool. In parts of zone 7 where snails and slugs are a problem, bare soil culture is preferred.

The robust foliage of celery or celeriac will cover the center area and shade out weeds, so cultivation is not necessary.

WATERING

Celery and celeriac are shallow-rooted crops, requiring a consistently good supply of soil moisture throughout their development. If no rain falls for 4 or 5 days, apply 1-2 inches of water. Furrow irrigation is preferred since sprinkler irrigation encourages the formation and spread of foliage diseases.

CONTAINER GROWING

Celery makes a fine, attractive container plant, but for the amount of vegetable harvested it takes up a lot of space. For this reason you should not attempt to grow celery in small containers. Rather, for best results, grow one plant per 3-5 gallon container and maintain a high level of soil moisture and fertility. It can reach to a surprising height and spread as much as 3 feet x 3 feet in soilless media, if fed frequently. For maximum yield from containers plant 3 stalks per container and harvest 2 of them to eat half grown. To harvest, slide a sharp knife under the root disc and avoid uprooting the plants that are to remain. Celeriac also grows well in containers. It can be crowded 4 inches apart in both directions if half the plants are pulled for eating when the roots are 1 inch in diameter.

ENVIRONMENTAL PROBLEMS

Perhaps the most common error in growing celeriac and celery is starting too late. The seeds sprout slowly and seedlings are slow to grow, so you must start seeds 10-12 weeks before the projected transplanting to the garden. With spring planted celery and celeriac there is a tendency to bolt. This occurs when the seedlings in the flat or the garden transplants are exposed to several days of cold weather (55° or below) when they are young. Being biennials, they react as if winter has come and gone, shooting up seed stalks as nature has programmed. Once bolting has begun, the process is irreversible and little of the celery plant will be edible. Other problems with celery and celeriac result from compacting of the soil around the roots. This can be avoided by using built up beds as described in SOIL PREPARATION.

BLANCHING

Individual celery plants can be blanched by covering them with a bushel basket for 7-10 days. Remove it during hot days or if humidity is causing the foliage to rot. To blanch a row of celery, stand 1x8 boards on edge up close to the plants, forcing the outer stems up. Gather soil from row middles to hold the boards up, or peg them in place. The boards will shut out enough light to blanch the stalks, yet the foliage canopy can spread out above it. Banking up with boards is a cleaner method than pulling soil or mulch up around the stalks. Also it helps to protect the lower parts of the stalk from fall freezes. The foliage may be frozen back but it gives temporary protection to the stems. Banking will blanch stems in 5-7 days, but inspect the plants frequently to see if they are rotting. If so, remove the boards.

HARVEST

Individual outer stalks may be cut as needed and new stalks will continue to develop from the center. To harvest a whole plant, cut just below the soil level. Pulling up the stalks is not recommended since soil can lodge in the heads. To harvest celeriac: The edible portion of celeriac is the knobby underground root or "bulb." When the root reaches a diameter of 2½-3½ inches, harvest by pulling the entire plant. Cut off and discard the top and the small branching roots.

STORAGE

Celery: Celery may be refrigerated in a moisture proof wrap for 1-2 weeks. Celeriac: Store celeriac root in moisture proof wrap in the refrigerator for up to 7 days.

STORING LEFTOVER SEEDS

When celery or celeriac seeds are stored as recommended, they should germinate at a rate of at least 50% after 3 years.

COLLARDS

Collards

Brassica oleracea Acephala Group, A "Cole Crop"

Cruciferae (Mustard Family)

Biennial, Grown As Cool Weather Annual

Full Sun Or Partial Shade

Rate Of Growth Rating, Direct Seeding -- 2; From Transplants -- 4

DESCRIPTION

Collards are easy to grow, and make a highly nutritious vegetable. They are more popular in the South and warm West where they provide fall and winter greens after most other vegetables except turnips have frozen out. Collards will grow well in every zone, but are relatively little known in the North, where their curly-leaved relative, kale, sells better. Mature collard plants are large, up to 3 feet tall and 2 feet broad, and relatively easy to harvest because the leaves are large and simple to wash.

Collards made a major contribution to the survival of southern frontier families because, when most other vegtables had frozen to the ground, families could always pick a "mess" of sweet, succulent collards for cooking with whatever meat or game was available or, in a pinch, feeding chickens or livestock.

Southerners seem to know intuitively when it is time to plant collards (usually around the first of August). The seedlings are heat resistant if given adequate water. Further north, collard seeds should be planted in the garden in early summer to grow into large, frost-sweetened plants before Thanksgiving.

A close relative of cabbage, collards probably originated in the same area in Europe.

SITE

In the North and Midwest, grow collard greens in full sun. In the South and warm West start seedlings in the shade made by spreading burlap over an arbor. You need to shade the seedlings only from afternoon sun. While not absolutely necessary, this procedure, when combined with frequent watering, will push collards along rapidly and result in earlier fall harvest. Don't plant collards where other cole crops have grown during the previous three years.

SOIL

Collards will grow on a wide range of soils, asking only to be supplied with sufficient water and plant food to maintain a steady, rapid rate of growth. Allowing collards to go dry can make the plants excessively fibrous and slow to recover when rainfall comes.

Collards are moderately tolerant of soil acidity, and will grow at pH ranges of 5.5 to 6.8. Over much of the South, ground limestone has to be applied every year or two to maintain the soil pH at levels that will support most vegetable crops. Collards are on the soil for a long time, and when picked heavily can extract large amounts of plant nutrients, especially nitrogen and potash, from the soil. Growing collards on very poor soil can result in restricted production, and foliage with excessive purple discoloration from phosphorus deficiency or brown-tinged from lack of potassium.

SOIL PREPARATION

Careful soil preparation is not necessary because collard seedlings are vigorous and strong growing. However, faster growth can result if the soil is prepared by working in manure or compost to spade depth and additional sources of nitrogen such as chicken manure or commercial nitrogen fertilizer.

PLANTS PER PERSON/YIELD

Grow 8 to 10 plants for each person to get 10 to 12 pounds of greens. Grow two to three times this quantity if you wish to freeze greens for winter use.

DAYS TO SEED GERMINATION

Soil Temp.	50°F.	60°F.	65°F.
Days to Germ.	14 days.	10 days	6 days

Soil Temp.	70°F.	75°F.	80°F.
Days to Germ.	5 days	4 days	3 days

Collard seeds are strong growing and will usually germinate 75 percent.

DIRECT SEEDING

In the South, "collard patches" are usually large, and direct seeding is preferred for the sake of economy. However, a hot, dry spell in late summer can delay emergence for several weeks and result in poor stands of small plants. The surest way to produce a successful crop of collards is to grow transplants, starting the seeds out-of-doors in a protected area four to six weeks prior to the indicated transplant dates. In all but zones 1 and 2, late summer planting is preferred because it lets the plants establish during warm weather then yield their edible leaves during the cool, frosty autumn season when flavor is at its peak.

When direct seeding, place three to four seeds every 12 to 15 inches, and thin to the strongest seedling in each location. Cover seeds ¼ inch deep.

TRANSPLANTING & THINNING

When transplanting collard seedlings in the South, it is best to wait until just after a rain during evening hours when the plants can adjust to their new location with minimum shock. In the arid, warm west, summer-transplanted collards have to be given shade. Often the seedling transplants are cut back to reduce the leaf area by about 30 to 50 percent in order to minimize the wilting which can occur during very hot, dry weather.

Thin seedlings or space transplants 12 inches to 15 inches apart in rows spaced 24 to 30 inches apart. You will save space by staggering the plants in double rows 18 to 20 inches apart.

SUPPORTS & STRUCTURES

Collard plants are strong and upstanding, with strong, thick central stems. They need no stakes.

RECOMMENDED MULCH/CULTIVATION

In the north, collards respond well to black plastic mulch which concentrates heat and promotes more rapid growth in summer and early fall. In the Deep South, plastic is not used for collards but, rather, deep organic mulches which keep the soil cooler and reduce evaporation.

Collards are deep-rooted, and one need not be extremely careful about cultivating around them. On heavy soils in the South they are often grown on ridged up beds, and cultivation consists of scraping soil out of the row middlings up around the plants to cover emerging weed seedlings.

WATER

Keep young transplants adequately watered. The soil should be moist but not water-logged. Collards need rain or irrigation every one to two weeks, depending on the water holding capacity of the soil. The smooth leaves shed water; sprinkler irrigation is okay.

CONTAINER GROWING

Collards are among the best plants for growing in containers for fall and winter harvest. The plants are large and should not be planted in containers of less than three to five gallons capacity per plant. An attractive display can be made with two or three plants grown in a shallow 7 to 10 gallon container. You can harvest enough collards for a small meal from these plants, and can get several pickings without harming the continued production of collards. Collards are one of the most winter-hardy vegetables you can grow in containers.

ENVIRONMENTAL PROBLEMS

Although collards are subject to the same diseases and insects as cabbage, they will usually survive and thrive because of their extreme vigor and large size. A frequent problem with collards is the assumption that they will make good summer greens. The crop isn't at its best for eating until cool weather comes.

COLLARDS

Another problem occurs in mild winter areas when collards have survived the winter. Some varieties will shoot to seed quickly, often as early as midwinter. By the correct choice of varieties you can minimize the problem of winter or early spring bolting.

HARVEST

Entire young plants can be pulled while thinning. The usual way to harvest is to start picking leaves from the bottom of the plant. If the bottom leaves become old and tough, skip over them and pick leaves higher up on the plant, leaving the growing point undisturbed. If you remove the central growing point by mistake, the collards will branch to a certain extent, but severe pruning delays the production of new leaves. If old, tough leaves of collards are harvested, the central rib should be removed and the leaf blades chopped in order to make them less chewy.

STORAGE

Collard greens can be kept in the refrigerator in a moisture proof bag 3 to 5 days.

FORCING FOR WINTER USE

In zones 1 through 4 collards can be over-wintered by watering the plants thoroughly, digging them with the roots on, harvesting the larger leaves for immediate use and transplanting the pruned plant to the protection of a coldframe. The plants will resume growth and will give you greens after extremely cold weather has stopped the growth of plants in the garden.

SEED STORAGE

Seeds placed in a sealed container with a desiccant, and stored in a refrigerator, should have a germination rate of at least 50 percent after 2-3 years of storage.

SWEET CORN

Sweet Corn

Zea Mays Var. rugosa (Zea mays Var. saccharata)

Warm Season Annual, Killed By Light Frosts

Full Sun

Ease Of Growth Rating -- 2

DESCRIPTION

Sweet corn is one the least efficient vegetables based on food production per unit of garden space and per unit of time. Yet, it ranks among the top ten in popularity because of its delicious taste when picked at peak maturity and cooked before a significant amount of kernel sugar converts to starch.

Even though you may never have grown sweet corn, you may have seen its sturdy plants, like tall grass. The tassels on top of the plant contain the male, pollen-producing flowers. Each silk on the ear is the visible part of a female flower, and can be pollinated by pollen from its own plant or by grains blown in from nearby corn plants. Plants can range in height from less than 4 feet to 8 feet or more, and average around 5 feet.

Sweet corn varieties differ in the height of stalks, number of basal tillers, ear size, foliage color, taste, tenderness, color and shape of kernels. Between varieties there is a range of more than 35 days to maturity, differences in the number of ears per plant, kernel depth and color, ratio of cob to kernels and ease of snapping the ears from the stalks.

The welter of choices could bewilder beginning gardeners were it not for screening by State Cooperative Extension Services to select varieties adapted to each region.

Most home gardeners who have the space, plant at least two varieties of sweet corn to spread the harvest over a longer period, and they plant in "blocks" rather than in single rows. This gives pollen from the tassels a better chance to fall on the silks. Pollination is necessary to the formation of the kernels.

Sweet corn can be grown successfully in any state in the Union except Alaska and also not across northern Canada. For many years it was seldom grown in the South because of foliage and root diseases, insects, and hot weather causing "pollen fire" at the time of pollen shedding. Now, however, adapted varieties are shifting southern tastes to sweet corn and away from "roasting ears".

Hybrids have nearly replaced the open pollinated varieties of sweet corn, and for good reasons . . . earliness combined with increased vigor, improved eating quality and disease resistance. Hybrids are also more uniform and concentrated in maturity. (The latter is not necessarily an advantage to home gardeners).

Sweet corn is native to the New World. Certain Indian tribes were planting sweet corn as well as grain corn when the pilgrims arrived but the colonists paid more attention to the field corn that could be stored and used in a number of ways. Sweet corn is not mentioned in literature until after the American Revolution.

SITE

Sweet corn needs full sun all day long and well drained soil. The late, more vigorous hybrids will tolerate heavy soil better than the extra early varieties. Good air movement should be considered in site selection; it can reduce the incidence of foliar diseases. Corn can cast shade on adjacent rows of other vegetables. Place corn rows on the north side of lower growing crops.

Rotate sweet corn crops to reduce the possibility of soil-transmitted diseases. If you plan to grow one of the supersweet hybrids that require isolation, check with your neighbors; they may be planting a standard variety just over the fence from your garden. Ask if they are planting popcorn seed. If so, specify sweet corn varieties that pollinate at least a week ahead of the popcorn in order to avoid crossing.

SOIL

Good crops of early sweet corn can be grown on sandy soil if it has an organic matter content of about 2%. However, maximum production of late summer corn can be achieved on medium to heavy soils which can hold a good reserve of soil moisture and plant nutrients. Very heavy soil should be worked up into raised beds or ridged rows to provide the good drainage that is essential to fast, healthy growth of sweet corn.

Sweet corn is moderately tolerant of acidity, and grows best at soil pH levels of 5.5 to 6.8. It is a moderately heavy feeder and a long-season crop. It shows nutrient deficiency symptoms clearly; thus, is used as an indicator for shortages of not only the major nutrients but also the secondary nutrients such as magnesium, sulfur, calcium and iron.

SOIL PREPARATION

Sweet corn is so demanding of good soil that experienced growers often begin preparing for corn crops the previous fall. They either grow a green manure crop for spading under the following spring, apply manure or, preferably, both. The growing crop will reduce the loss of nutrients from the manure and the residual manure will help to break down the green matter when it is turned under.

Soil for sweet corn should be deeply dug and organic matter incorporated throughout the root zone to a 12" depth. If coarse organic matter is used the surface of the seed bed should be raked to remove any chunks which might interfere with germination.

PLANTS PER PERSON/YIELD

Fifteen to twenty plants is about the smallest amount that is practical to grow in a block for good pollination, and this presents a problem. Modern hybrids will keep on the stalk for 7 to 10 days at the most, with the exception of the EH hybrids. A family of two would have to stuff themselves with sweet corn to keep up with production. Therefore, if you grow corn, plan to freeze or can some of it to reduce the waste. Planting smaller blocks isn't the answer, poor pollination can result.

The yield from 15 to 20 plants will range from 15 to 40 useable ears, depending on the variety, planting time and condition of the soil. While two or three ears of small varieties might be needed per serving, a single ear of very large, late hybrids will sometimes suffice.

SEED GERMINATION RATE

Soil Temp.	41°F.	50°F.	59°F.	68°F.
Days to Germ.	No Germ.	22 days	12 days	7 days

Soil Temp.	77°F.	86°F.	95°F.	104°F.
Days to Germ.	5 days	4 days	4 days	No Germ.

DIRECT SEEDING

Sweet corn seed is normally direct sown. Early-sown seeds often rot in the ground before they have a chance to sprout. This can be caused by cool, damp soil conditions and poor drainage, and can be diminished by treating seeds with a fungicide seed protectant purchased in a garden center. Sowings made later in the season, after the soil has warmed, are less affected by seed rot.

Corn seed can be sown in a number of ways. The most common arrangement is in rows 2½ to 3 feet apart, with the seed 4 to 6 inches apart in the rows. Cover seeds with 1 inch of soil and later thin plants to stand 8 to 12 inches apart. Single, high-ridged rows tend to dry out too rapidly in arid climates.

Another approach is planting in hills — placing 4 to 5 seeds in groups with 2 inches between seeds. Space the hills 2 to 3 feet apart. Thin to 2 to 3 plants per hill. Planting in hills is usually preferred when you are growing a relatively small number of corn plants. The reference to "hills" comes from building up low mounds of soil for planting each group of seeds and "hilling up" the soil around plants.

You can also grow sweet corn in double rows in a built-up bed, four feet wide. Sweet corn seed is often "listed" (planted in the bottom of deep furrows) in windy, dry climates such as the Great Plains. The furrow protects the sweet corn seedlings until they are 4 to 6 inches high and betters their chance of survival once they have grown above the protection of the furrow.

Always plant corn seed in blocks of at least four rows or hills side by side rather than in one or two long rows. Planting in blocks assures adequate pollination which is essential for the development of fully filled ears.

It is occasionally difficult to get a good stand of corn when seeding in late summer when the soil is quite warm and dry. Under such conditions, dig deep furrows, flood them two or three times with water to wet the subsoil to a depth of 12 inches or more, plant the corn in the bottom of the furrow and cover with dry soil. The reserve moisture should be sufficient to germinate the seeds; if none have sprouted within a week, flood the furrow again.

When planting corn in sticky, clay soil, crusting sometimes causes seed emergence problems. To avoid crusting, try covering with sand or compost.

TRANSPLANTING AND THINNING

Except in Zones 1 and 2, little is gained by transplanting corn. In cold climates you can get a jump on the season and have an early corn harvest by using transplants. Corn can be transplanted successfully if seedlings are grown in plantable pots and if the root system is not disturbed at planting time. Sow seeds indoors four weeks before the indicated direct seeding date; plant two seeds per pot. Keep the temperature for sprouting as close as possible to 75°F. until shoots show, then gradually lower the temperature to the mid-sixties. Grow the young seedlings in full sun or under a fluorescent light. After seedlings have grown to a height of 2 to 3 inches, clip off the weaker seedling in each pot. Harden the young plants for at least ten days before setting them in the garden. Be prepared to protect early transplantings against frost.

Corn plants should stand 8 to 12 inches apart in rows, but extra early varieties can be spaced as close together as 6 inches in fertile soil. Generally, the later and more vigorous the variety, the more space you should allow between plants. If you are planting a lot of corn on moderately fertile soil, and will not have time to "pamper" it, you may wish to increase the spacing somewhat between plants.

Plants grown in hills can be thinned to 8 to 12 inches apart. The population of corn plants for a given area should be about the same whether planted in hills or single or double rows. Remember that spacing rows too close together can cause the tall corn plants to shade themselves.

Crowding corn plants can reduce yield and the size of ears. Thin excess corn seedlings when they are about 4 inches tall and preferably when soil is moist. Excess plants can be pulled easily without disturbing adjacent seedlings.

MULCHING AND CULTIVATION

Black plastic works well as a corn mulch over most of the country. In zones 5, 6 and 7 it is advantageous to keep the soil cool for optimum growth of late-planted corn crops. There, organic mulches work better; they combine well with wide beds because the mulch can be spread between the double rows of corn, leaving the pathways open for drainage and/or irrigation.

If corn is not mulched, it should be cultivated two or three times during the life of the crop to control weeds. Early cultivations can be fairly deep, but should not be done close to corn plants. The depth of cultivation should decrease as the crop grows, so that surface roots will not be destroyed. Commercial growers often pull up soil from the "middles" between rows and deposit it around the base of the plants. This practice is particularly helpful where corn is grown on ridged-up rows and subject to the soil eroding away from the roots because of heavy rains.

The use of clear plastic mulch in zones 1 and 2, applied two to three weeks prior to planting time to warm the soil can bring in corn crops a week or two earlier than those planted with black plastic mulch. The clear plastic will admit and retain more solar energy.

SUPPORTING STRUCTURES

Well grown corn plants are self-supporting. Usually, if stalks are knocked over by high winds during rain storms, you can set them upright and keep them standing by throwing a spade full of soil on the uprooted side.

WATERING

Sweet corn has a high requirement for soil moisture, especially during the period beginning ten days before the appearance of silk and continuing 5 to 12 days after the silking date. At other times, corn needs to be irrigated every 10 to 14 days if no rain falls. Apply 3 to 6 inches of water to wet the soil to a depth of 15 to 30 inches. Such heavy applications would be wasted on fast draining, sandy soils. They should be given about 2 inches of water per application and watered weekly. Sweet corn should be furrow irrigated to avoid wetting the foliage.

CONTAINER GROWING

Early, compact varieties of sweet corn make a decorative but not very practical container plant. You can grow about six stalks in a 30-gallon container and, if you assist the plant during the pollen shedding stage by shaking pollen to fall on silks, you can get a fairly good fill of kernels and plump ears. However, even with the most productive varieties, your maximum crop will be 5 to 10 small ears, and you have to ask yourself if this amount of work and expense would not be more productive if applied to an adaptable vegetable.

ENVIRONMENTAL PROBLEMS

One of the most common problems with sweet corn is the failure of the ears to fill out completely. This results in what are called "nubbins" and can be

SWEET CORN

due to failure of sufficient pollen to fall on receptive silks. If you are planting in blocks, look for other causes. Nutrient deficiencies, chiefly of available phosphorus, water shortages at critical growth periods, too high temperatures, and too thick planting can interfere with pollination. In northern states, prolonged wet, cool weather can cause broken rows of kernels and nubbin ears.

Phosphorus is most likely to be deficient on acid soils in high-rainfall areas and in arid areas where high soil pH levels can transform the phosphorus into compounds that are not available to plant roots.

HARVESTING

When the silks first show dry brown tips, the ears are usually ready for harvest. Check the maturity by stripping an ear partially and piercing a few kernels with your thumbnail. The skin should be tender and the juice that runs out should look "milky" — not clear (a sign of immaturity) nor pasty (a sign of overmaturity).If the corn is not ready, wait a day or two and check again.

Harvest by snapping off individual ears with a sharp twist. Some varieties such as *Iochief* are hard to snap. Cook sweet corn immediately after harvest to enjoy its full sweetness; otherwise, husk the ears at once and put them in plastic bags in the refrigerator until you want to cook them.

STORING LEFTOVER SEEDS

When stored with a desiccant in a sealed container and placed in a refrigerator, sweet corn seeds should germinate at least 50% after three years.

STORAGE

Picked sweet corn deteriorates very rapidly at room temperature. To keep 1 or 2 days, keep the corn cool and moist by refrigerating in a moisture proof bag.

CUCUMBER

Cucumber

Cucumis sativus, The Green American, European And Middle Eastern Types And Lemon Cucumber

Cucumis melo (Flexuosus Group) Armenian Cucumber

Cucurbitaceae (Gourd Family)

Warmth-loving Annual, Killed By Light Frost

Full Sun And Free Air Movement

Ease Of Growth Rating, Garden Cucumbers -- 5, Greenhouse Cucumbers -- 10

DESCRIPTION

The fruits of cucumbers are not highly nutritious, but are so refreshing and delicious that they have been a favorite in salads and for pickling for thousands of years.

Cucumbers grow on large-leaved vines which range in length from two feet in dwarf types, to five or six feet in standard varieties. The vines cling weakly by tendrils but do not twine, and need frequent training when grown up supports to save space.

Until recently, the "sex life" of cucumbers was uncomplicated because each cucumber vine had both male and female blossoms. Cucumber fruits develop only from female blossoms and, by nature's plan, only when pollen is transferred from the male to the female blossom by insects. All of this was changed when, in 1950, breeding stock for a new "gynoecious" cucumber was introduced to North America from Korea by the USDA. Gynoecious cucumbers have all female or mostly female blossoms and have the potential for greater production than open pollinated types. Seeds of gynoecious varieties are usually sold in packets which include a few seeds of standard open pollinated varieties to supply male blossoms and pollen for fruit development. More recently, "parthenocarpic" cucumbers have been developed; these do not require pollination to form fruits, and are virtually seedless. Some of the very newest varieties are both gynoecious and parthenocarpic.

During the past forty years plant breeders have concentrated on improving cucumber varieties. A textbook for college students in horticulture, circa 1939, reported "There are very few varieties of cucumbers grown in the United States . . ." and went on to mention only ten varieties. Since that time, plant breeding has focused on multiple disease resistance, better pickling and storage (shipping) qualities and on manipulating sex expression.

Plant breeders also incorporated the "burpless" characteristic from so-called oriental varieties and are now creating short-vined hybrids for space efficiency and container growing. Belatedly, they have recognized the superior eating quality of the non-bitter Middle Eastern cucumbers with smooth, tender skins. These fruits do not require peeling. Genes for disease resistance are being incorporated into the Middle Eastern cucumbers to combine their excellent taste, bitter-free characteristics and liberation from the chore of peeling, with the American resistance to the many diseases which plague cucumbers.

Cucumbers belong to the same plant family as muskmelons, watermelons and squash, and are in the same genus as muskmelons. Crosses between these genera have been attempted by plant breeders but have not been successful to date. Therefore, don't worry about these vegetables crossing under your garden conditions.

Cucumbers are probably native to Asia and Africa. Evidence exists of their cultivation in western Asia for at least 3000 years. Distinct species of cucumbers, one of them used more for cooking than for salads, have been grown in India for as long as recorded history. Ancient Greeks and Romans introduced cucumbers into Europe, and seeds were brought to North America by the early colonists. In 1539, records indicate that American Indians in Florida were growing cucumbers, perhaps descended from seeds brought with the settlers of St. Augustine.

SITE

Plant cucumbers where they can receive 6 to 8 hours of sun and free air movement all day. Under dry climate conditions, especially where strong wind is a factor, cucumber vines can be battered and stressed by winds unless adequately shielded by wind-breaks. Elsewhere, avoid pockets of still air which encourage high humidity and foliage diseases.

Although cucumbers prefer full sun all day, they can adapt to morning or afternoon shade if they receive direct sun the balance of the day.

In zones 6 and 7 it is difficult, but not impossible, to grow cucumbers in midsummer because of extreme heat and/or prevalence of foliage diseases.

SOIL

Cucumbers grow best on clay loam or clay soil with reasonably good drainage. On sandy soil in warm climates, cucumbers tend to mature quickly and usually only one or two pickings are realized. However, with ample moisture and nutrients, many pickings can be made.

Cucumbers will tolerate reasonably acid soils with a pH as low as 5.5, but grow best in the 6.0 to 6.5 range. Cucumbers are moderate to heavy feeders. Gardeners aim for sustained heavy production. Toward that end, organic plant foods can be as effective in maintaining vigorous vines as periodic applications of manufactured fertilizers.

SOIL PREPARATION

No other crop, with the possible exception of melons, responds as well to the addition of manure to the soil. In years past, farmers would dig a hole, mix in a bushel of rotted manure with the excavated soil, and use it to refill the hole. This would create a low mound or "hill" for planting seeds. The decomposing manure and slight elevation would create good drainage and an excellent rooting area. Be advised that a layer of unmixed, coarse manure or leaves beneath soil can also create a barrier to the capillary movement of water from lower layers of soil up to the surface. It is best to incorporate organic matter thoroughly into the soil and plant your seeds on the resulting raised area.

Sandy soil can be improved by incorporating large quantities of manure or other organic matter which, after decomposing to humus, will increase the moisture holding and nutrient storage capacity of the soil.

Cucumbers have a high requirement for potassium and nitrogen, and a moderate need for phosphate.

PLANTS PER PERSON/YIELD

Grow 3 to 4 plants for each person to get 2 to 3 dozen slicing cucumbers or 3 to 4 dozen picklers. To produce this quantity, you will need to keep your vines healthy and productive throughout 6 to 8 pickings.

SEED GERMINATION RATE

Cucumber seeds are strong growing and long lived. Low soil temperatures can delay germination and increase the risk of seeds rotting in the soil or seedlings being killed by disease organisms either before or shortly after they emerge. High soil temperatures can produce weakened seedlings, but this is usually a consideration only in zones 6 and 7 during midsummer. Midsummer planting can be successful in western areas where ocean breezes or fog tempers the summer temperatures.

Soil Temp.	50°F.	59°F.	68°F.	77°F.
Days to Germ.	No Germ.	16 days	7 days	5 days

Soil Temp.	86°F.	95°F.	104°F.
Days to Germ.	4 days	3 days	2 days

DIRECT SEEDING

Rushing the planting season isn't worthwhile because plants stunted from cold weather never catch up in size or productivity with plants not exposed to cold. In zones 1 and 2, where late springs and cool summers can delay the planting and maturity of cucumbers, you can build an inexpensive solar trap to make the soil warm up faster. Two weeks prior to the frost free dates, cover a raised seedbed with a sheet of clear plastic and bury the edges in a trench. Two weeks later, cut slits and plant seeds. The trapped heat will hasten germination and early growth.

Cucumber seeds can be "drilled" in straight rows 4 to 6 inches apart or planted in hills, 3 to 4 seeds per hill. Seeds are usually drilled when being run up a trellis or arbor, when a large crop is being grown for pickling. Hill planting is used when vines are being run up teepees or when only a few plants are needed.

Cover seeds ½ inch deep in heavy soils and 1 inch deep in sandy soils. Germination can be improved in heavy soils by planting seeds in shallow furrows and covering them with ½ inch of sand.

Plantings made in late summer in hot, dry soil are usually covered to a depth of 1 inch with porous material such as sifted compost. In areas where hot winds make midsummer seed germination difficult, cucumbers are planted by "listing", digging a furrow to a depth of 6 to 8 inches, flooding it with water, and sowing seeds in the bottom of it (Dryland farmers use listing for planting grains in dry soil).

TRANSPLANTING AND THINNING

The earliest varieties of cucumbers can be transplanted. Start from seeds sown indoors 4 to 5 weeks prior to the indicated direct seeding times. Sow 3 to 4 seeds per peat pot, covering ½ inch deep. Keep them as close as possible to 70°F for good germination. Make sure seedlings are in full sun as they grow. Cut off all but the two strongest seedlings. Harden transplants and be prepared to protect them from late cold snaps.

Space transplants or thin seedlings to two plants per location. The non-vining, bush types can be spaced 7 inches apart in rows 24 inches apart. The vigorous vining types should be spaced 7 inches apart in rows 36 to 48 inches apart.

MULCHING AND CULTIVATION

Cucumbers respond positively to mulching. In zones 1 and 2, use clear plastic mulch; in other areas use black plastic. In zones 5, 6 and 7, the soil can become too hot in the summer for the use of plastic mulches, and organic mulches are preferred for successive crops. If you use organic mulches for your spring crop, wait until the soil has warmed before pulling it in under the vines. Straw makes a good mulch for cucumbers when applied to a depth of 4 to 6 inches, but be sure to scatter some nitrogen fertilizer before applying the straw.

Without mulching, cucumber vines will need to be turned back to enable you to scrape out weeds with a hoe. Do this carefully so as not to break the vines, and don't try moving them after the vines have grown more than two feet in length. If you can't get straw, hay or leaves for mulching at this stage, use flattened cardboard boxes or several layers of newspapers.

SUPPORTING STRUCTURES

Lemon cucumbers and the standard pickling or slicing types don't require support, but they are often trained on uprights as a means of conserving garden space. Support is necessary for production of perfectly formed cucumbers of the long fruited burpless types and European greenhouse types, as well as Armenian cucumbers. If allowed to crawl across the ground, these types produce curved, distorted fruits.

A four foot tall support is adequate for lemon and pickling types; others should be provided strong, sturdy supports at least six feet tall. Cucumbers climb by weak tendrils, and bearing vines are heavy; use strong twine, wire or a cage structure such as a cylinder of fencing or reinforcing mesh. A teepee can be made by lashing together three or four rough poles or stakes. Most cucumbers resist climbing early in the season, so you'll have to pamper stems by training them up the support and tying them occasionally.

WATERING

Cucumbers require a consistently high level of soil moisture, especially during the fruit development stage. They need 2 to 4 inches of rain or irrigation every seven to ten days, depending upon the moisture holding capacity of the soil. Lack of adequate moisture can cause bitterness and also result in nutrient deficiencies resulting in deformed fruit.

CONTAINER GROWING

The small-vined cucumbers developed for intensive gardening or growing in containers are variously described as "midget", "dwarf", "compact", or "bush". Your best bet are the varieties with vines that range from 18 to 24 inches in length; these fall into the "compact" class. Vines of this length are more productive than the smaller midget or dwarf types which sacrifice some production for the small sized plants. Short-vined cucumbers should be grown in containers of 3 to 5 gallon size, one to two vines per container. Larger containers of ten to thirty gallon capacity can be used to grow three to five vines of standard varieties.

ENVIRONMENTAL PROBLEMS

Bitterness in cucumbers is occasionally a problem. It is an inherited characteristic which is made worse by stresses such as high temperatures and inadequate soil moisture. If you encounter bitterness in cucumber fruits, it may help to peel them and remove ½ inch of fruit from the stem end where the bitterness is concentrated. An even better idea is to switch to one of the non-bitter varieties.

Malformed fruits occasionally show up in cucumbers and are due mostly to inadequate pollination or dry soil. Crooked fruits can also signal a deficiency of phosphate. Cucumbers will occasionally fail to set fruits for extended periods of time due to a lack of pollinating insects. Don't assume that bees have been killed by spraying; it may be that a string of dark days has discouraged blossoms from opening fully so that they can attract bees. You can overcome poor pollination in trellised vines by transferring pollen from male to female blossoms with a camels hair brush. You should not become concerned if the first blossoms are all male; this is typical of all but the gynoecious types of cucumbers; after a week or two of all-male blossoms you will begin to see female blossoms appearing.

You may occasionally see a "shift in sex" of gynoecious cucumbers due to adverse weather or other stresses such as lack of water. The plants may be growing normally and producing all female blossoms as they should, but when a period of adverse weather sets in, you may see abnormally large numbers of male blossoms on supposedly gynoecious plants. Once the weather stress is over, the plants will revert to their all female or near female habit.

HARVEST

The best stage for harvest depends on the fruit type and intended use. Any type can be harvested for salad use at any size from baby fruits with the flower still attached, all the way through maturity when the fruits start to lose their color and quality. Harvesting fruits at small stages reduces the total yield. Generally, mature or nearly mature fruits are the best for brining because the flesh is firm and the interior solid.

Once production starts, harvest cucumbers frequently so more will form; check plants daily. Clip or carefully twist off fruits to avoid breaking the vines.

STORAGE

Cucumbers may be kept fresh in a moisture proof plastic bag in the refrigerator for one week. However, cucumbers for pickling *must* be freshly picked, not over 24 hours old.

SEED STORAGE

Seeds placed in a sealed container with a desiccant and stored in a refrigerator should have a germination rate of at least 50 percent after 3-5 years. In general, cucumber seeds hold their germination quite well.

CUCUMBER

EGGPLANT

Eggplant

Aubergine, Guinea Squash

Solanum melongena

Solanaceae (Nightshade Family)

Warm Season

Perennial, Grown As An Annual

Damaged By Prolonged Cool Weather

Full Sun

Ease Of Growth Rating, From Purchased Plants -- 4; Grown From Seedlings -- 6

DESCRIPTION

Eggplant fruits are produced on large, downy-leaved plants which occasionally develop small spines on their stems and on the caps of the fruit. Plants of modern varieties average 2 feet high and 2 feet wide, but older varieties may grow twice that size where summers are long and warm.

In the past eggplant was difficult to grow, since it matured late and set fruit sparsely during cool summers. In zone 1, eggplant consistently failed, and in zones 2 and 3 one was fortunate to harvest 3 or 4 fruits per plant. Then, plant breeders developed earlier varieties that would fruit reliably in northern zones and resist the major diseases. However, they were concerned most with the needs of commercial vegetable shippers, so they developed new varieties with thick, tough skins. These tended to taste bitter when picked at full maturity.

The breeders then utilized the tender skinned Oriental varieties to develop early bearing, non bitter hybrids which resist some of the major diseases. These early bearing varieties can produce a dozen or more fruits per compact, bushy, plant.

Authorities disagree on the origin of eggplant, but it appears to be native to the Indian subcontinent, and cultivated for many centuries in China, India and Arabia. Eggplant was probably introduced to Europe during the Moorish invasion of Spain.

SITE

Eggplant requires full sun all day. Only in zones 6 and 7 is afternoon shade appreciated. In zone 2 and windy areas along the West Coast, windbreaks will speed early growth. Avoid planting where related crops such as tomatoes, peppers and potatoes have been grown during the previous 3 years.

SOIL

Eggplant is most productive on well drained, medium to slightly heavy soil. Deep sandy soils make it difficult to maintain high levels of moisture and nutrients eggplant requires. Grown on sandy loam soils, eggplant will bear 5 to 7 days earlier than the same variety planted on heavy clay soil. It is moderately tolerant of soil acidity and will grow at pH ranges of 5.5 to 6.8. In more alkaline soils good drainage is necessary to keep salts from accumulating.

SOIL PREPARATION

Built up beds will provide the warmer soil and improved drainage that eggplants appreciate. Incorporate moderate to heavy amounts of organic matter and an application of preplant fertilizer. If heavy amounts of manure or commercial fertilizer are used, give the beds a thorough watering and wait 2 to 3 days before setting in transplants; this reduces the chance that salts will build up and danger of root damage if hot, dry weather closely follows transplanting.

PLANTS PER PERSON/YIELD

Plant conservatively, allowing 2 to 3 plants per person. This should yield 20 to 24 fruits of the large-fruited types, or 50 to 65 fruits of the small to medium sized Oriental types.

SEED GERMINATION

Eggplant seeds need warm soil to germinate and are usually started indoors. They germinate within a narrow range of temperatures. The optimum soil temperature for germination is 85°F. Light improves germination.

Soil Temp.	68°F.	77°F.	86°F.
Days to Germ.	13 days	8 days	5 days

The usual germination percentage is approximately 70%.

TRANSPLANTING AND THINNING

Sow seeds indoors 8 to 10 weeks before the transplant date. Plant seeds ¼ inch deep, about 4 seeds per inch. When seedlings are 2 inches tall, thin or shift to individual pots. Harden the plants before setting them in the garden.

Eggplant seedlings are sensitive to cold. Consequently, transplanting should be delayed until soil has warmed considerably and for 2 to 3 weeks after all danger of frost has passed. Cloche or hot cap coverings are recommended, especially in cool mountain, coastal, and northern areas.

In zones 6 and 7, eggplant seeds can be started for the second crop by planting them in a warm protected corner of a well watered nursery bed. Trim off 1/3 of the foliage before transplanting. Set transplants 18-24″ apart in rows spaced 36-48″ apart, depending on the variety. In zones 5, 6, and 7, where eggplants grow quite large, allow more space between plants in the row.

Eggplant seedlings are difficult to grow indoors because of their high requirement for heat, and even more difficult to transplant to the garden without checking their growth. Yet, experienced home gardeners do it routinely by treating eggplant seedlings as if they were tropical plants. They use bottom heat for starting seeds at 75 to 80°F and grow the seedlings under fluorescent lights at 70-75°. They transplant late, 7 to 10 days after tomatoes are set in the garden, and always shelter the young plants with bottomless plastic jugs or tunnels of clear plastic. The protection is more against wind injury than against cold. All these safeguards are to continue the lush, rapid growth they made in the warm environment indoors, without growth checks. The object is not to grow plants so that they form blossoms when quite small but, rather, when they have developed large frames capable of supporting heavy crops of eggplant fruit.

RECOMMENDED MULCH/CULTIVATION

In zone 2 and in high elevation areas of zones 3 and 4, eggplant responds well to mulching with clear plastic. Clear plastic traps solar heat and encourages fast growth, which makes the production of eggplant practical where it otherwise might not be.

Black plastic mulch is standard for eggplants in other zones, but decomposed pine needles or hardwood leaves also make an excellent mulch if applied after the soil has warmed. Make the layer of organic mulch at least 2 to 3 inches deep to shade out weeds and to keep the soil moisture at a consistently high level.

Eggplant is deep-rooted and can be hoed or kept clean between rows with a tiller set for shallow cultivation.

SUPPORTING STRUCTURES

Where summers are long and moderately hot, eggplants can reach 4 feet or more in height. Support is essential to prevent branches from breaking as they become laden with fruits.

A low cage made of heavy wire fencing or mesh is the simplest device. Make a cylinder 18 to 20 inches in diameter and place it around the plant early in the season, or drive 3 or 4 stakes a foot into the ground, 6 inches from the base of the plant, and lace with twine to hold the plants erect. Eggplants can be espaliered on small, fan-shaped trellises. These are especially attractive for container grown eggplants.

WATERING

Eggplants are fairly drought resistant, but letting them go without rain or irrigation for more than 7-10 days will stress the plants, reduce production, and may cause malformed fruits. Furrow or drip irrigation is preferred; sprinklers can encourage foliage diseases. Apply 3 to 4 inches of water to wet the soil to a depth of 15 to 20 inches.

CONTAINER GROWING

Any eggplant variety will grow well in containers, but varieties with smaller plants, such as Dusky and Early Black Egg, are more manageable. Provide at least 3 gallons of soil medium for each plant. A shallow, 10 gallon tub will hold 3 plants. The contrast between the gleaming purple fruit and the greenish-gray foliage makes eggplant an attractive patio plant.

Near the end of the season, cover eggplants in containers with clear plastic to prolong fruiting and set the containers under an overhang to ward off frost as long as possible.

ENVIRONMENTAL PROBLEMS

A common problem with eggplants is the failure to set fruit, or setting only a light crop. This is usually caused by a poor choice of variety. Many garden centers still sell seed and plants of the old, late-maturing varieties which will not set fruit reliably during cool summers or when stressed by weather.

Another problem with eggplant is bitterness. It usually appears about the time seeds start to turn dark within the fruits. Bitterness can be avoided by picking fruits at a younger stage of growth or by specifying a variety which has been bred to be non-bitter at large sizes.

HARVEST

Standard types of eggplants are at their prime when the fruit diameter is 3 to 4 inches and its length is 6 to 7 inches. The Oriental types should be grown to the same length, but they will be much more slender. For both types, the skin should have a high gloss at harvest; a brownish cast is a sign of over maturity. The skin should not spring back if pressed with the thumb; in the cut fruits, seeds should be barely noticeable and not dark. Harvest all mature fruit promptly to prolong production.

Eggplant stems are tough; harvest by cutting them with pruning shears or a sharp knife. Handle carefully to avoid being pricked by spines at the stem end of the fruit.

STORAGE

Eggplant keeps 1 week refrigerated in a moisture proof bag.

STORING LEFTOVER SEEDS

When stored as recommended, eggplants seeds should have a germination rate of at least 50% after 3 to 5 years of storage.

ENDIVE

(Escarole, Curly Endive, Broadleaved Endive, scarole)

Cichorium endivia

Compositae (Composite Family)

Cool Season

Biennial, Grown As An Annual

Killed By Heavy Frosts

Full Sun Or Partial Shade

Ease Of Growth Rating -- 3

DESCRIPTION

Endive is a rather large, low-growing, curly or wavy-leaved plant grown mostly for salad greens, but occasionally as a potherb. Escarole or scarole is a separate class of endive with less frilly leaves and loose heads which blanch somewhat better at the center. Both types can spread to cover an 18 inch circle. Of the two, the curled or fringed is more popular, but mostly because of its more attractive appearance on produce shelves and salad bars.

Actually the wavy-leaved escarole has a notably better quality and is more tender when blanched. In Europe red-tinged, semi-heading are also popular.

Endive harvested during the summer can be bitter. The flavor is much improved when seeds are planted for harvest after the weather has turned cool. The flavor is then sweet and the leaves are tender.

Probably of East Indian origin, endive was used as a food by the Egyptians at a very early date. It was mentioned by Pliny as being eaten both as a salad and as a potherb.

SITE

Plant endive in a sunny area. However, where summers are extremely hot, keep your seedlings under burlap or lath shade until they are half grown. Gradually remove the shade when the main heat of the summer is over. The plants will have gained considerable growth over seeds planted in full sun. Another method is to start seedlings in flats or cold frames in a lightly shaded area, and transplant them into the garden once the intense heat has lessened.

SOIL

Endive will grow under a wide range of soil conditions from 5.0-6.8 pH, and in more alkaline soils if drainage is good and the crop is watered frequently. It grows best on moderately fast draining clay loam or muck, but will do well on sandy loam if it is supplied with enough organic matter to maintain an even level of soil moisture and plant nutrients.

SOIL PREPARATION

Since endive is usually planted in midsummer for maturity in the fall and early winter, provide for good drainage around the maturing plants. Endive and escarole do poorly when water is permitted to stand around the crown of maturing plants for any length of time. Rotting will occur and spread rapidly throughout the head.

Endive grows best when planted on a wide bed. Plant 2 rows of endive per 4 foot wide bed, setting the seedlings near the shoulders. Cut a trench down the center of the bed and use the excavated soil to build up ridges. Leaving the endive plants on ridges 4-6 inches above the surrounding soil will keep the crowns dry and greatly reduce the chance of basal rot. In arid climates you can use the deep central furrow to hold water for irrigation. When heavy rains fall, open both ends of the furrow for it to serve as a drainage ditch.

PLANTS PER PERSON/YIELD

Grow 6-8 mature plants for each person. This will yield 6-8 pounds.

SEED GERMINATION

Endive seeds will germinate at soil temperatures of 45-75°F but the optimum range is 60-65°F. Prolonged exposure of endive seeds and seedlings to soil temperatures in the 40°F range can trigger premature bolting. A typical lot of seeds will germinate 70-75%.

DIRECT SEEDING

Hot weather causes bitterness in endive, even when it is properly blanched. Schedule planting so the crop reaches full development during the cool weather of late spring and early summer or during the fall. In the zones 6 and 7 it can be grown for winter harvest also. Established endive is fairly frost tolerant and can be left in the garden until the onset of severe cold.

Endive grows quickly and easily from seeds sown in the garden. Sow seeds at a rate of 6-8 seeds per foot and cover only 1/8-1/4 inch deep. Make rows 18 inches apart. Endive is a suitable crop for planting in wide rows. Plant 2 rows per 4 foot wide bed, sow seeds 2 inches apart. Begin harvesting whole plants are still small, with leaves 3-4 inches in length.

TRANSPLANTING AND THINNING

Sow seeds indoors approximately 4 or 5 weeks prior to the indicated transplant dates and harden young plants before setting them in the garden. Some gardeners prefer to grow endive from transplants for their spring gardens. Having well-grown endive seedlings ready for transplanting at the frost-free date in zones 1 and 2 avoids the risk of bolting that can come with early direct seeding in the garden.

Mature plants should stand 18 inches apart in the row to develop fully. They may be cramped, but some crowding can be tolerated. Thin several times to even out the stand but don't begin until seedlings are 2 inches tall. They can be carefully moved to other spots in the garden where they will mature slightly later, thus staggering the harvest.

MULCHING AND CULTIVATING

Mulching endive or escarole with organic mulch such as straw helps to keep foliage clean, to conserve moisture, and insulate the soil. Begin pulling mulch up around the plants when they are as large as your hand.

Before mulching however, make sure the ground slopes away from crowns of the plants in all directions or the mulch will hold too much water around the plants, encouraging basal rot. Mulching will keep the ground from freezing in the fall and will prolong the growth of endive beyond its normal span. In the event of a hard freeze early in the season, straw mulch can be pulled up over the plants for temporary protection.

The seedlings grow slower than most summer garden weeds and must be hand weeded when young. Later on, the spreading plants tend to smother out the weeds growing close to them, but you will need to hoe out the survivors. Endive is deep rooted, and you need not be overly careful when hoeing around the plants.

WATERING

Endive has a deep taproot, and will survive long drought periods, but the quality of the leaves will suffer. For best quality, endive needs rain or 2-4 inches of irrigation every 7-10 days. However, excessive moisture standing around the plants can cause crown rot.

CONTAINER GROWING

Endive makes a fine container plant. Individual plants can be grown in containers as small as 1 gallon in capacity.

ENVIRONMENTAL PROBLEMS

Bitterness is a problem with endive, and there are no "non-bitter" varieties. The bitterness is always present but is less noticeable as the weather turns cooler, and particularly if the endive is blanched. (See BLANCHING).

Bolting is a problem when endive is grown in mild winter areas and harvested through the winter. Some or all plants may bolt if warm spells follow prolonged cold weather. Once bolting starts, there is nothing you can do to stop it. Harvest endive immediately at the first evidence of bolting.

BLANCHING

A simple method of blanching is to space plants closely, forcing the leaves to grow more erect. But a more complete blanching can be achieved by covering individual plants with large pots or baskets or by tying the tips of outer leaves together with a rubber band.

HARVESTING

Harvest the entire plant after it has been blanched for a few weeks to improve the flavor. If endive is not blanched, begin harvesting when the center leaves begin to show a creamy heart. Because of their bitterness, you may wish to discard the outer leaves, retaining only the creamy or blanched center.

Another method is to dig the plants, roots and all, after a thorough watering and set them in cold frames. Snap off the outer leaves; the plants will regrow, providing greens for weeks after garden plants have frozen.

STORAGE

Store endive wrapped in moisture proof wrap in the refrigerator up to five days. Endive is not suited for canning, freezing, or drying. Plants can be moved from the garden to a cool cellar where they will keep 3-5 weeks, blanching slowly during storage.

STORING LEFTOVER SEEDS

When stored as recommended endive seeds should have a germination rate of at least 50% after 3-5 years.

KALE

Kale

***Brassica oleracea* Acephala Group**

Cruciferae (Mustard Family) A "Cole Crop"

Biennial, Grown As A Cool Weather Annual

Frost Hardy; Killed By Prolonged Freezing Weather

Full Sun Or Light Shade

Ease Of Growth Rating, From Direct Seeding -- 2;

Transplants -- 3

DESCRIPTION

Kale is one of the most productive and nutritious of all vegetables, and attractive as well, yet is down the list in popularity. This is probably because it is principally a potherb and not everyone likes cooked greens. You can use the small leaves of kale in salads but the flavor is a bit heavy and the texture chewy for some tastes.

Some restaurants line salad bowls with leaves of young kale because it will not wilt as fast as leaf lettuce.

Kale looks like erect, open growing, non heading cabbage but its leaves quickly identify it even when planted side by side with other close relatives such as collards. New leaves continue to form as the mature kale leaves are harvested. Only a few plants of kale can give a family meal after meal and a surplus for freezing. Harvest can commence in less than two months from seeding and will continue well into winter.

Kale was known to the ancient Greeks and was grown by the colonists. It should not be confused with Chinese kale, a smooth-leaved relative grown for its central flowering shoot. Modern cultivars of kale are short and compact, with large, broad leaves that simplify harvesting and washing.

SITE

Kale prefers full sun in zones 1 through 3. Further south and west, light afternoon shade keeps the hot sun of late summer from bleaching the leaves. Kale is a cole crop and should not be planted where related crops have been grown during the previous three years.

SOIL

Provide a fertile soil, moisture-retentive but not soggy. Good drainage will reduce the risk of soil borne diseases and will prolong production if the fall season is wet. Kale is a heavy feeder, drawing on soil reserves of all three major nutrients, especially nitrogen. It needs consistently fertile soil to maintain high yields of well-colored leaves. Kale will grow well at soil pH levels of 5.5 to 6.8, and beyond these limits if no serious nutrient deficiencies occur and levels of organic matter are high.

SOIL PREPARATION

Quick sprouting, strong growing kale will succeed on rather shallow and roughly prepared soil. Yet, there is little reason for such haphazard treatment. Kale usually follows a quick spring crop such as mustard greens or leaf lettuce, therefore working up a fine seedbed is easy. Work in well decomposed organic matter but make sure it does not include trimmings from cole crops.

The plants of kale are so broad that they are usually grown in single rows; in high rainfall areas the rows are usually ridged up.

PLANTS PER PERSON/YIELD

Grow 8 to 10 plants for each person, if you like kale and plan to freeze the surplus. You should harvest a minimum of 10 to 12 pounds from this number of plants.

SEED GERMINATION

Kale seed is strong and fast growing. The usual germination is 75 to 80 percent. It will sprout in 5 to 7 days at temperature ranges of 50 to 90°F but, being a cool weather crop, prefers soil temperatures of 60 to 70°.

DIRECTIONS FOR DIRECT SEEDING

The most satisfactory plantings of kale are those timed for harvest during cool or cold weather. Kale will grow and produce during the heat of summer, except in the south, but its flavor and texture suffer from the heat. Direct seed by placing a pinch of 2 or 3 seeds where you want a plant to grow. Allow 12 to 18 inches between plants. When sowing during the summer, plant a bit deeper than the usual ¼ inch. Water frequently so that the seedbed doesn't dry out. Light shade on the seedbed is helpful.

TRANSPLANTING AND THINNING

Kale is rarely transplanted, except in zones 6 and 7 where high soil temperatures in late summer can impede seed germination. There, seedlings can be started in a shaded area out-of-doors and, when ready for transplanting, watered and trimmed to reduce the foliage area by 30 percent. These precautions, plus planting during evening hours and watering frequently until the plants have "taken", should ease their transition to the hot, windy garden environment.

Thin direct seeded kale to leave only the strongest seedling from the pinch of seeds at each point. Surplus seedlings can be transplanted. Use the 18 inch spacing between plants unless you garden intensively and water and feed seedlings frequently.

MULCHING AND CULTIVATION

Organic mulches help keep the soil moist and cool in late summer. You can draw it up under the kale plants to control weeds. Plastic mulches aren't appropriate as they tend to keep the soil warm. In the western part of zones 6 and 7, clean culture is usually preferred because slugs, snails, earwigs and sowbugs breed under organic mulches.

Cultivate kale plants by scraping off the weeds with a hoe. Nothing is gained by hilling up soil around plants. Kale seedlings are easy to distinguish from most weeds.

SUPPORTS AND STRUCTURES

Kale plants are self-supporting. Only the old fashioned tall 'Siberian' kale may require staking.

WATERING

Kale has large smooth leaves that use a great deal of water in transpiration, especially during dry, windy weather. Irrigation or rain at least weekly is recommended to keep it flourishing and in good color. Wet the soil to a depth of 12 inches by applying 2 to 3 inches of water by furrow irrigation or sprinkler.

CONTAINER GROWING

Garden kale makes a beautiful and productive container plant. A container of 7 gallon capacity can hold three plants and these can supply enough leaves at one harvest to feed a small family. Harvest only the outer leaves, retaining the inner leaves to support the plant while it grows replacement foliage.

Use ornamental kale if you are more interested in decoration than eating qualities.

ENVIRONMENTAL PROBLEMS

The most common complaint with kale concerns tough, "blah tasting" cooked leaves. This can always be connected with spring-seeded kale, harvested during hot weather. Generally, kale has many of the same insect and disease problems as cabbage, but is less affected because of its large vigorous plants.

HARVESTING

If you are like most gardeners, your kale will get ahead of your needs. You will tend to leave the older outer leaves and pick the smaller ones, nearer to the top of the plant. This will work, but be sure not to take the growing tip because it will delay the formation of additional leaves by two to three weeks. The older leaves may be a bit tough but they have the maximum concentration of nutrients and are fine for cooking when chopped. Remove the central rib.

STORAGE

Kale can be kept in the refrigerator in a moisture proof bag 3 to 5 days.

STORING LEFT-OVER SEEDS

Kale seeds sealed in an airtight container with a desiccant and placed in a refrigerator should keep with little loss in germination for three years, providing the seed was relatively fresh when the package was opened.

KOHLRABI

Kohlrabi (Turnip Cabbage, Stem Turnip, Turnip-Rooted Cabbage)

***Brassica oleracea* Gongylodes Group (Formerly *B. oleracea* Var. *caulo-rapa*)**

Cruciferae (Mustard Family), A "Cole Crop"

Biennial, Grown As A Cool Weather Annual. Killed By Heavy Frosts

Full Sun Or Light Shade

Ease Of Growth Rating -- 3

DESCRIPTION

Kohlrabi is grown for its small, cabbage-like plants which have globe-shaped, swollen stems just above the ground. These swellings are often referred to incorrectly as "bulbs". No one has come up with a more appropriate name such as the term "hips" used for the fruits of rose bushes.

Kohlrabi does not taste like cabbage or turnips, but has a mild, sweet taste all its own. The leaves can be eaten but should be chopped and the stringy

stems discarded. The leaves can be cooked alone or combined with the peeled roots.

Kohlrabi can grow to softball size but is best eaten at a diameter of 2 to 3 inches. Plant seeds thickly and eat the young plants which you pull out during thinning. You can regulate the maximum size of the stems somewhat by close spacing but an even better approach is to make successive plantings of short rows to maintain production of tender stems in the preferred 2 inch diameter range.

Kohlrabi is thought to have been derived from wild cabbage during the 16th century, and originated in northern Europe. Its name is derived from the German words "kohl" (cabbage) and "rabi" (turnip). This vegetable oddity is not mentioned in ancient writings. It can be trimmed and stored for several weeks in a cool place, so it can be surmised that the history of kohlrabi paralleled the development of root cellars. Prior to the advent of canning, Northern Europeans were so dependent on root cellars to carry their families through 6 to 7 months of cold weather that any relief from a steady diet of stored cabbages, turnips and rutabagas would have been welcome.

SITE

Grow kohlrabi in full sun. It will tolerate light shade for a few hours daily, but will grow rather tall and spindly. In zones 6 & 7, afternoon shade lowers the soil temperature and makes earlier seedings of fall crops possible.

SOIL

Any soil except fine sand is suitable for kohlrabi, but a moist, fertile soil is preferred. The object is to grow it rapidly in order to prevent the swollen stems from becoming fibrous; therefore, it helps to modify the soil with plenty of organic matter for storing nutrients and moisture. A soil pH range of 6.5 to 7.5 is preferred.

SOIL PREPARATION

For spring crops in zones 3 to 7, raised beds are recommended because the soil will warm up faster and will push the kohlrabi stems to harvest size before hot weather. In zones 1 & 2, except on heavy, poorly drained soils, planting "on the flat" gives as good results. For fall crops you may wish to plant on the flat because raised beds are difficult to water except by sprinkling.

PLANTS PER PERSON/YIELD

Grow 12 to 16 plants for each person to get 4 to 5 lbs. of kohlrabi. Grow less if you are just testing this delicious vegetable.

SEED GERMINATION RATE

Kohlrabi seeds are comparable to cabbage in germination performance. Seeds will germinate in 5 days at 70°F.

DIRECT SEEDING

Kohlrabi is usually direct seeded. It will emerge reliably and grow quickly in the seedling stages. Plant seeds in rows 12 to 18 inches apart, with 2 seeds per inch, or in wide bands with seeds broadcast 1 inch apart. Cover seeds ¼ inch deep. Use sand to cover spring plantings as it will encourage faster germination.

TRANSPLANTING AND THINNING

Transplants offer little advantage but are sometimes used for the earliest crop. Sow seeds indoors 4 to 5 weeks prior to the direct seeding dates and harden young plants for 10 to 14 days before shifting to the garden.

While seedlings are still small, thin to stand 3 to 4 inches apart in rows or 4 to 5 inches apart in wide bands. Begin harvesting when stems have swollen to 1 inch diameter to make room for the remaining plants to expand.

MULCHING AND CULTIVATION

Kohlrabi grows so rapidly that mulching is not necessary for spring crops. However, for crops planted in late summer for fall harvest, an organic mulch will insure a steady supply of soil moisture which promotes rapid growth and tender stems. Weeds compete strongly for water and nutrients and will need to be pulled out by hand from between the closely spaced kohlrabi.

WATERING

Kohlrabi can withstand short periods of dry weather but such stresses should be avoided if you want tender, fiber-free stems. If no rain for 5 to 7 days, apply 1 to 2 inches of water to wet the soil to a depth of 5 to 10 inches.

CONTAINER GROWING

Kohlrabi makes an excellent, attractive container crop because it can be crowded. Thinned out plants can be eaten for greens or small roots. You can plant seeds in containers as early as leaf lettuce, and, with this early start, you are assured of harvesting tender, full-sized roots in late spring. In a shallow tub of 5 to 7 gallon capacity, you should be able to grow up to a dozen kohlrabi.

ENVIRONMENTAL PROBLEMS

Fibrous stems are the major complaint with kohlrabi. Fiber is present in all varieties, but more noticeable in the older standards. Weather stresses and drought bring out this characteristic, so it is most important to force kohlrabi to quick maturity by maintaining high levels of soil moisture and plant nutrients.

Spindle-shaped roots are another problem; these can result from crowding or from inadequate or irregular soil moisture. If occasional misshapen roots are seen, it could be due to poor quality control by the seed producer.

HARVESTING

Pull entire plants, beginning when stems have swollen to 1 inch diameter. The plants are strongly anchored and will pull easier if the soil is moist. Try to harvest all stems at less than 3 inches in diameter, when the incidence of fiber is low.

STORAGE

Kohlrabi bulbs can be stored in a moisture proof bag in the refrigerator for up to 2 weeks. The leaves can be stored in a moisture proof bag in the refrigerator for 1-2 days. Plants will keep in the garden for a few weeks after frost if covered with straw or dry leaves. Trimmed stems will store in a root cellar and don't give off a strong odor like turnips.

STORING LEFTOVER SEEDS

The longevity of kohlrabi seeds is similar to that of cabbage.

KOHLRABI

LEEKS

Leeks

Allium ampeloprasum **Porrum Group**

Amaryllidacea (Amaryllis Family)

Biennial, Grown As A Long Season, Frost Hardy Annual

Leeks Prefer Full Sun All Day But, In Zones 5, 6 & 7, They Benefit From Light Afternoon Shade

Ease Of Growth Rating, Direct Seeded -- 4; From Home Grown Transplants -- 5

DESCRIPTION

Leeks are related to onions and somewhat resemble giant scallions. Well grown, fully mature leeks have long, tender white stems which can be as much as 2 inches in diameter by 9 inches long. The long, flattened, silvery-green leaves are tough and are seldom eaten. The white stems make fabulous soups and are used in stews and salads as well.

Leek plants can reach 12 to 18 inches in height and spread, but can be crowded in the row to produce a great deal of valuable food in a small area.

Leeks are popular in Europe but little known in the U.S.A., perhaps because they suffer from the mistaken reputation of being difficult to grow. If you can meet four important requirements, you can produce big, tender leeks:

Loose, fine-textured, fertile soil, high in organic matter.

Proper timing of planting.

Planting in trenches or drawing soil up to the stems for blanching.

Raised beds if your soil is slow to drain.

Leeks have been cultivated in their native Mediterranean area since ancient days.

SITE

Place leeks to the side or back of the garden so you will not have to walk around them to reach your salad crops. Don't grow leeks where other onion family members were grown the previous year.

SOIL

Leeks are only slightly tolerant of soil acidity and will grow best at pH levels of 6.0 to 6.8. They are moderately heavy feeders but have small root systems and are not good foragers. Plant food should be placed close to the plants. Grow leeks in deeply worked soil, finely pulverized and high in organic matter.

SOIL PREPARATION

Dig the soil to a depth of at least 12 inches, breaking up clods and removing roots and debris. Incorporate well-decomposed organic matter such as sifted compost or moistened peatmoss, and pre-plant fertilizer. If your garden soil drains slowly, excavate soil from paths to build up leek beds to stand 4 to 6 inches above the surrounding area. Cut trenches 4 inches deep and 24 inches apart for the leek seeds or transplants; leave the ends open for drainage. The trenches will make it easier for you to pull up or "bank" soil around the stems as they grow. The 24 inch spacing will give you an area to borrow soil for hilling without stripping off surface roots.

On sandy or well-drained soils or in low rainfall areas, raised beds and trenches aren't necessary. Hilling up or banking will cover stems adequately to blanch them.

Deep, mellow, fertile soil will stimulate the fast growth necessary to produce big leeks.

PLANTS PER PERSON/YIELD

For fresh use grow 10 to 12 mature leeks per person. This will yield 2 to 3 lbs. trimmed. Leeks store well; grow up to 2 dozen plants per person and store the surplus in the garden under a mulch, or in a root cellar.

DAYS TO GERMINATION

Leek seeds will germinate at soil temperatures as low as 50°F, but slowly. At 70°, seeds will germinate in 7 to 12 days.

DIRECTIONS FOR DIRECT SEEDING

Space seeds ½ inch apart; cover to a depth of ¼ inch with porous soil or sand. When planting in late summer cover seeds ½ inch deep in furrows that have been flooded to wet the soil 6 to 12 inches deep. Sifted compost makes an excellent seed covering.

TRANSPLANTING AND THINNING

Leek seedlings are rarely available for purchase; plan on growing your own. To get pencil-sized seedlings for transplanting, start seeds indoors for spring transplanting 10 to 12 weeks before the frost free date. The seedlings grow slowly. Sprout at temperatures of 60-70° F and grow seedlings in a cool area. Harden off for 3 to 4 days before setting them in the garden. Set plants 4 to 6 inches apart.

Seedlings for summer transplanting can be started out-of-doors in a flat of potting soil, 8 to 10 weeks in advance. Sprout seeds in an area protected from strong sunlight and drying winds. Trim back the tops to reduce shock at transplanting. Water transplants twice daily for a week.

Thin direct-seeded leeks before they reach pencil size and eat the surplus. For maximum size, grow leeks 6 inches apart; under intensive culture, 4 inches apart.

MULCHING AND CULTIVATING

Clear plastic mulch will hasten the maturity of spring-seeded leeks in zones 1 through 3. Apply after hilling-up. You will need to shade out the weeds that will grow in the "greenhouse" formed by the plastic. In the summer, pile organic mulch on top of the clear plastic; it will shade out the weeds and can be pulled up around the leeks to blanch the lower stems. Use organic mulch alone for summer planted leeks; it will keep the soil cool and conserve moisture. Pile such materials as straw, hay, shredded leaves or composted sawdust to a depth of 4 to 6 inches around the leeks.

Leek seedlings closely resemble grass. For this reason, seed leeks in straight rows so you can distinguish them from weeds.

WATERING

Leeks are a shallow rooted crop and need rain or irrigation at least every week throughout the season. Apply 2 to 3 inches of water to wet the soil to a depth of 10 to 15 inches. Either sprinkler or furrow irrigation will do.

CONTAINER GROWING

Leek makes a fine container crop because it can be crowded. The excess plants can be pulled out of the porous soil without disturbing adjacent plants. At pencil-size the seedlings make good substitutes for scallions.

Containers for leeks should be at least 18 inches deep and filled only halfway before seeding or transplanting. As the seedlings grow, fill in around them with planter mix, leaving about one inch of headspace for watering.

ENVIRONMENTAL PROBLEMS

Leeks are reasonably easy to grow if you start them at the recommended season for your zone and grow them in fertile soil with plenty of water and plant food.

The most common complaint is failure of leeks to size up before hot weather stops their growth or kills the plants. If you garden in a cool summer zone and this is your problem, choose an early variety, plant earlier and hustle the plants along with plenty of plant food and water.

In zones 6 & 7 where leeks are commonly grown for winter harvest a common problem is plants bolting to seed before they have reached eating size. This is due to planting too late. Despite having to deal with very hot soil in late summer, you must plant leeks sufficiently early or they will go into the winter at a small size. Cold soil will keep them from growing much until spring comes and along with it, the longer days that bring about flowering.

HARVESTING

Leeks mature so slowly that home gardeners usually begin harvesting them as soon as the stems are pencil sized, leaving just enough plants to protect under a fall mulch in the garden or to store in a cool root cellar for winter use. Stems with a diameter of 1 to 2 inches store best: they begin to firm up at a diameter of ¾ inches. Loosening leeks with a spading fork expedites pulling and reduces bruising of stems. Trim off all but two to three inches of leaves when storing inside; leave as much of the root system as possible. Set the plants upright in clean moist sand. At temperatures of 32 to 40°F leeks will keep for several weeks.

STORAGE

To store leeks: Refrigerate in moisture-vapor proof bags and use within 7 days. Where the ground does not freeze, leeks can remain in the ground throughout the winter and be dug up when needed. In cold winter areas, leave in ground by covering with straw over black plastic.

STORING LEFTOVER SEEDS

When placed in an airtight container with a desiccant and stored in a refrigerator, seeds should keep for a year. Leek, onion and related seeds are poor keepers.

LETTUCE

Lettuce

Lactuca sativa

Compositae (Composite Family)

Cool Season Annual, Killed By Heavy Frost

Full Sun Or Partial Shade

Ease Of Growth Rating, Direct Seeded -- 2; Transplanted -- 4; Crisphead -- 5

DESCRIPTION

Lettuce is the most popular salad crop, and is grown for its succulent foliage. It is emphatically a cool weather vegetable. The plants are small and, with the exception of crisphead varieties, matures so rapidly that they are space-efficient and valuable.

Most lettuce varieties fit into one of four classes: loose leaf, crisphead, butterhead and cos or romaine. Loose leaf, butterhead and cos varieties are customarily direct seeded; crisphead lettuce varieties are usually transplanted.

Some kinds of lettuce are easy to grow, others are not. Part of the secret is in the timing of planting; part is in providing fertile soil and a consistent soil moisture level, but part is in choosing the right class of lettuce for your climate and your degree of gardening skill.

Loose leaf lettuce is the easiest to grow, providing you skip the midsummer months. But, even if you succeed in growing it during the summer heat without its going to seed, the taste is likely to be bitter, especially with the bronze or red varieties. Far northern, high-altitude and cool coastal states are exceptions; with the aid of shading from slats or burlap you can harvest fairly good lettuce in midsummer.

Leaf lettuce seeds are generally seeded direct in the garden beginning as early as the soil can be worked, and at 2 week intervals through mid-spring. Seeding halts then, and resumes in late summer, or even later in zones 6 and 7.

The Crisphead, Iceberg or Great Lakes type heads you see in the stores are usually shipped in. They are more difficult to grow and contain less vitamins than leaf lettuce. Being later maturing, they are not usually direct seeded. Start seeds indoors and transplant well grown, hardened off seedlings to the garden in mid-spring and again in late summer. Your chances of success are greater with late summer or fall plantings of crisphead lettuce, because lettuce sends up flowering shoots on lengthening days. Lettuce for fall harvest will not bolt except during extended periods of warm, bright winter weather in southern and warm western areas.

The butterhead varieties are increasing rapidly in popularity because they are almost as easy to grow as the loose leaf varieties, have a high content of vitamin A, and a distinct flavor, especially when the centers or "hearts" (actually more like rosettes) are blanched to a cream color. Butterhead varieties mature early enough to be direct seeded early in the spring, but the best and sweetest crops are grown from late summer direct seedings.

Cos or romaine lettuce is one of the Cinderellas of the lettuce world, waiting to be discovered by home gardeners. It develops somewhat slower than the loose leaf varieties and is recommended for late summer or, in mild climates, fall planting. Maturing in the fall or later, the upright, cylindrical, loosely wrapped heads separate easily into cupped leaves and creamy, crunchy centers. The heads are rather large and, if you need only a few leaves, you can take them off the outside of the head, leaving the center leaves to grow.

In parts of Asia, lettuce is used more for cooking than in salads. Their loose leaf types are preferred for lettuce soup.

The first mention of lettuce as a salad crop was a record of its appearance on the tables of Persian kings in 550 B.C. One could speculate whether this was due to the succulence of lettuce or the presence of the mild narcotic, lechuga, in the milky juice of certain wild species. Lettuce has little caloric value and the value of its vitamins and minerals was unknown in those days. It was not the fare of common people, but early colonial records indicate lettuce was planted in the gardens of the early French forts in the Great Lakes area.

SITE

Lettuce, except for crisphead, grows so rapidly that it is often tucked in among other slower growing plants in the garden, or among summer flowers. Spring lettuce beds are harvested early enough to be used for warm season crops, thus increasing the value of your garden space.

Most gardeners prefer to grow three or four varieties for different colors and textures in salads, and keep a steady supply by planting small amounts of seed at two week intervals skipping, of course, the summer months.

Lettuce prefers full sun in the spring and fall, but welcomes light shade where it can be grown in midsummer.

SOIL

Lettuce is only slightly tolerant of soil acidity, and grows best at a pH range of 6.0 to 6.8. It requires light amounts of nitrogen and phosphorus, and moderate levels of potash. For spring crops of lettuce, light, fast-draining soil is preferred because it will warm up quickly and will consistently release higher levels of nutrients and water to the plants than heavier soils. If you have the option, make late summer plantings on heavier soils because these do not lose nutrients to leaching as rapidly as the sands. Only in areas where fall rains keep the soil too wet should lettuce beds be raised for late summer plantings.

SOIL PREPARATION

Spring lettuce beds are often prepared in the fall and the surface of the soil left rough or covered with poultry manure. As soon as the soil can be prepared in the spring, the seedbed is raked fine and lettuce seeds planted. On all but sandy or well drained loamy soils, spring lettuce benefits from the improved drainage and faster warmup provided by raised beds or ridged rows. Lettuce will not tolerate salty soils. In western areas salty soils are usually leached before planting, by ponding 6 to 12 inches of water in a basin and letting it dissolve and carry the salt to deep layers in the soil. Late summer or fall plantings of lettuce usually follow an earlier vegetable, and respond to having a modest amount of well decomposed organic matter worked into the soil prior to planting.

PLANTS PER PERSON/YIELD

Crisphead type — grow 4 to 6 plants for each person to get 6 to 8 pounds. Other lettuce types — grow 8 to 10 plants for each person to harvest 4 to 5 pounds. Remember to plant lettuce in small increments because there is not much you can do with a surplus except give it away or throw it on the compost heap.

SEED GERMINATION

Lettuce seed germinates strongly; in fact, the federal minimum standard for germination is 80%. It has two peculiarities: it will not germinate when the soil is extremely hot, and it needs light for optimum germination. From the latter, you might gather that lettuce seed should be sown on top of the soil, but if you do so it will grow poorly because of alternating wetting and drying. Consequently, the best way to plant lettuce seed is to cover it lightly with a porous material so the strong sunlight can penetrate just enough to meet the requirements of the seed.

DAYS TO GERMINATION

Soil Temp.	32°F.	41°F.	50°F.	59°F.
Days to Germ.	49 days	15 days	7 days	4 days

Soil Temp.	68°F.	77°F.	86°F.	95°F.
Days to Germ.	3 days	2 days	2 days	No Germ.

From this data, you can see that lettuce seeds will germinate in very cold soil, but emergence takes so long that the seeds and seedlings are exposed to rotting due to attacks by soil pathogens.

DIRECT SEEDING

Direct seeding can be used for late summer, fall and winter planting of all types of lettuce and for spring plantings of looseleaf and butterhead types.

Lettuce seeds will germinate in cold soil, but seeding is usually delayed until the soil temperature has exceeded 40° in order to maximize germination and survival. Young seedlings will tolerate light frost. During cool weather, cover seeds to a depth of

LETTUCE

only ⅛ inches. Seed sowings made in the summer, or later, can be covered to a depth of ¼ inch.

Plant about three lettuce seeds per inch and space them carefully so as not to have them coming up in clumps. Late plantings for autumn and winter crops require extra care to prevent drying of the seedbed. Sow seeds deeper than for spring planting and protect the seedbed with light shade. Burlap placed on top of the seedbed will retain moisture and keep the soil cool. Sprinkle water over the burlap and remove it when the first seeds sprout. Don't wait too long or the seedlings will become entangled in the burlap.

On heavy clay soils, rains can cause a crust to form and "seal in" the seeds. On such soils, plant lettuce seeds in shallow furrows and cover with a porous material such as vermiculite or finely sifted compost or sand.

TRANSPLANTING AND THINNING

All lettuce can be handled as transplants, but this method is used mainly for spring plantings of cos and crisphead types. Four to six weeks prior to the indicated transplant time, sow seeds indoors. Cover seeds to a depth of ⅛ inch. Lettuce seeds need full sun or strong light from fluorescent tubes to germinate. Plants need to be grown in a cool area to keep them from becoming leggy. The transplants are fragile and should be hardened off for several days before planting in the garden.

Protective coverings should be used to shield transplants from wind and from late frosts.

Direct seeded looseleaf and butterhead types can be thinned to a spacing of 3 to 4 inches. Use the later thinnings for salads. If you seed too thickly, or if the seedlings come up in clumps, your first thinning should be made when seedlings are only 2 to 3 inches tall. You may have to pull out and discard small clumps of lettuce plants rather than individual seedlings.

Transplanted looseleaf and butterhead types are usually spaced 5 to 8 inches apart. Make rows 12 to 18 inches apart. Looseleaf and crisphead lettuce are "naturals" for wide band planting and for interplanting among slower growing vegetables or flowers.

Space cos lettuce 9 to 12 inches apart in rows 15 to 18 inches apart.

Crisphead types require the most space, usually 12 inches between plants and 18 to 24 inches between rows. Of all the lettuces, crisphead types are harmed most by crowding in the row.

MULCHING AND CULTIVATION

Lettuce plants love organic mulches, but do not apply them on spring crops which need warm soil to bring them quickly into maturity. Reserve organic mulches for crops planted after mid-summer for fall or winter harvest. Plastic mulches are rarely used because lettuce is in the ground for only a short time.

Lettuce has a shallow root system, and is generally hand weeded. Weeds can be scraped from around larger plants, but you should be careful not to scatter soil on the foliage.

SUPPORTING STRUCTURES

Lettuce plants are self-supporting and do not need structures.

WATERING

After direct seeding or transplanting keep the soil moist but not waterlogged. If you have a dry spring, you may have to sprinkle seeded areas once or twice daily to bring up seeds quickly. When plants are established, avoid excessive watering because it can cause leaf rot and, in crisphead types, puffy heads. Furrow or drip irrigation is generally preferred for leaf lettuce. Sprinklers can splash soil on leaves and heads.

Lettuce wilts quickly when water levels in the soil begin to drop. If no rain falls for four to five days, apply one to two inches of water by irrigation. This should wet the soil to a depth of 5 to 10 inches.

CONTAINER GROWING

Lettuce is well adapted to containers, and grows rapidly in porous, fast warming artificial growing media. Experienced container growers usually transplant lettuce seedlings around the base of other container plants, thus getting a double crop from the same space. Leaf lettuce plants can be crowded in early stages and the excess plants removed and eaten.

A one gallon container will support two to three plants of looseleaf lettuce, one crisphead lettuce, or one to two plants of butterhead or cos. However, containers of three to five gallon sizes are best suited for growing lettuce because they neither dry out as fast nor blow over like small containers.

ENVIRONMENTAL PROBLEMS

Three common problems with lettuce are planting too much, too late and too close. If your lettuce has in previous years gone to seed or turned bitter from summer heat you have seeded too late. Direct seed a small area early in the spring when it is still cold. You will be pleasantly surprised by how well your lettuce will grow during the early spring months, and how the harvest period is prolonged.

Gardeners in zones 5, 6 and 7 have discovered lettuce seeds are difficult to germinate in late summer due to extremely hot soil. This failure to germinate is caused partly by a built-in mechanism that inhibits lettuce sprouting in hot soil. You can neutralize this mechanism by sprouting lettuce seeds before you plant them. Mix seeds with pre-moistened peat moss, enclose in a bag and put the mixture in the refrigerator for a week. Then, place the bag in a room at 65 to 70 degrees F. in indirect sunlight. Turn the bag every day and watch for signs of sprouting. Do not put the bag in strong direct sunlight.

At the first sign of sprouting roots or shoots, sprinkle the peat moss, seeds and all, in the bottom of a furrow which has been presoaked by ponding water in it. Cover very lightly with vermiculite or sifted compost, neither of which will get as hot as sand or soil. Lightly sprinkle two or three times daily until the seedlings have emerged, then reduce the frequency to every few days.

Gardeners soon learn that except in cool summer areas it is useless to direct-seed crisphead, cos and some butterhead varieties out-of-doors in early spring. Heads will not mature reliably prior to summer. These types are much better suited to planting in late summer.

Gardeners who live where the spring season is short, often prefer to use the heat resistant variety Salad Bowl, as it will give them one to two weeks of additional harvest prior to bolting.

Lettuce Mosaic is a virus disease which occasionally affects garden lettuce but is most troublesome in commercial crops. This disease can be carried in the seeds themselves, and will show up in the crop as mottled leaves and malformed plants. The disease can also be introduced and spread by insects.

Look for the initials "MT", for mosaic tested, on seed packets and in catalogs. This symbol indicates that the seeds have been tested and are substantially free of seed-borne mosaic virus.

HARVEST

All lettuce passes its prime quickly. Have succession or relay plantings coming along to provide a steady harvest. Pulling lettuce, roots and all, can get soil on the leaves and complicate washing; slice or snip off plants just above the soil level.

You can harvest looseleaf lettuce at any time the leaves are large enough to eat, but harvest and washing are simplified when you wait until the leaves are about the size of your hand. You can take individual outer leaves or harvest the entire plant. Pull the leaves before any yellowing occurs.

Harvest crisphead lettuce by cutting the entire head when it feels firm. Don't wait until the outer leaves begin to turn color. Crisphead varieties are concentrated in maturity, so you may wish to harvest the first heads just as the inner leaves are beginning to fold over.

The central leaves of butterhead form a small, soft head which is harvested by cutting the entire plant. You may wish to harvest a few outer leaves one at a time.

Harvest the entire plant of romaine or cos lettuce or the outside leaves only. This type forms a moderately firm, cylindrical head of upright leaves.

STORAGE

Lettuce can be kept refrigerated in a moisture vapor proof bag for 5 days. Store away from other vegetables and fruits to prevent russet spotting. A covered lettuce crisper may be used. Lettuce for salads can not be frozen, canned or dried.

FORCING FOR WINTER USE

Lettuce is one of the best crops for growing under plastic canopies, cloche or coldframes for winter use. Lettuce can grow well in the reduced sunlight in late fall. Harvests from lettuce grown under protective structures can extend the production season by as much as four to six weeks.

STORING LEFTOVER SEEDS

Lettuce seeds are fairly long lived. Seeds placed in a sealed container with a desiccant and placed in a refrigerator should germinate at least 50%.

MELONS

Melons

(Cantaloupe Or Muskmelon And Related Melons: Casaba, Crenshaw, Honeydew, Persian, But Not Watermelon)

Cucumis melo Reticulatus Group. Cantaloupe Or Muskmelon And The Persian Melon.

Cucumis melo Inodorus Group. Casaba, Crenshaw And Honeydew

Cucurbitaceae (Gourd Family)

Warm Season Annual Killed By Light Frosts

Damaged By Prolonged Cool Weather

Full Sun

Ease Of Growth Rating -- 5

DESCRIPTION

Until a few years ago, melons were grown mostly in medium to large gardens because of their sprawling vines, low production per sq. ft. and long growing season. Now, you can buy earlier hybrids that will yield up to 6 or 8 fruits per plant on shorter vines.

Melons are not yet a sure-fire crop for beginners but are worth experimenting with until you get the hang of it. Home garden culture is virtually the only access to the exquisite flavor of vine-ripened fruit from the gourmet varieties. By and large, the top-quality gourmet varieties of cantaloupes produce fruits that are too large or too soft for long-distance shipping. You may find them being grown by market gardeners for local sale.

Melons have been subjected to a concentrated improvement program since the early 1900s. Early efforts focused on developing varieties that would mature in short-season areas. Then, for years, breeders focused on varieties for long distance shipping. Recently, with the advent of hybrid melons and a better understanding of the inheritance of disease resistance, breeders turned again to home garden varieties to introduce superb flavor, multiple disease resistance, high production under adverse weather, and holding power after peak harvest. These newer melons are significantly superior to the old standards, particularly in the cantaloupe or muskmelon class.

Most recently, breeding projects are yielding earlier maturing varieties in the Inodorus Group that will perform under humid conditions. Formerly, it was a feat to grow casaba, crenshaw or honeydew melons outside of the long, warm, dry regions of zones 6 and 7. Foliage diseases and a tendency of the fruit to split during warm, wet weather are still serious drawbacks to growing these melons east of the Mississippi River but, with a little extra care, it can be done.

Cantaloupes or muskmelons and the related melons look much alike in plant habit. Most are standard vine types, with robust runners that can spread widely. The leaves are large, with rough surfaces and rounded lobes. Male and female blossoms are produced on the same plant and bees and other insects are necessary to transport pollen from male to female blossoms. The fruits are usually hidden by the foliage canopy while small, but become more visible as they near full size. The temptation to pick fruits at this stage is great but should be resisted. Melons need several warm days to sugar-up after they reach full size.

Melon seeds and seedlings are sensitive to cool soil or cold air. They are one of the last vegetables to be planted, to give the soil a chance to warm to at least 65° F. Vines need full sun all day and good air movement.

Although some authorities hold that melons originated in Persia, others maintain that their home is in Africa. A small fruited variety has long been grown throughout the country of Malawi. Traders could have taken seeds from interior Africa to the Middle East. In history an intriguing mention is made that "Columbus found melons growing on Isabella Island in 1494." No wild melons survive to authenticate their origin. Melons were grown in Virginia in 1609 and along the Hudson River in 1629.

In gardening history, this was a late start, but melons were so admired that they quickly reached a position of prominence in the warm, dry belt stretching from Morocco to China and Japan, and across the USA and southern Canada. Prior to 1870 melons were seldom seen on American markets. The introduction of the small-fruited, thick-rind, Netted Gem variety in 1881 launched the melon shipping industry.

SOIL

Only slightly tolerant of acid soil, melons grow best at a pH range of 6.0 to 6.8. However, they are routinely grown in the arid West at pH levels up to 7.5, where careful watch is kept on the availability of major elements and micronutrients. Incorporating manure or compost into high pH soil buffers it and decreases the effects of excessive salinity. Melons like sandy loam or lightweight soil underlain with a clay base. The latter is ideal, as it combines warm soil with good drainage and a reservoir of soil moisture. You can grow good melons on heavy soil but they will mature a few days later than on lighter soils.

SOIL PREPARATION

For maximum production from sandy soils, amend them generously with manure or decomposed compost. On heavier soils incorporate organic matter but, as well, build raised beds or ridged up rows. If water stands for some time after a rain, consider building frames of wood or cinder block and filling them with a porous mixture of equal parts of sand, compost or manure, and pulverized garden soil. Frame culture can make a great deal of difference in the rate of growth, earliness and total production on heavy, wet soils.

PLANTS PER PERSON/YIELD

Production from melon plants is variable and depends not only on the variety planted but also on length of season, freedom from diseases and a consistently good level of nutrients and water in the soil. Typically, 2 to 4 healthy plants should produce from 8 to 16 fruits but certain recently introduced hybrids can produce considerably more. Production of melons tends to be concentrated, with several of them ripening at once. Where the length of season permits, make relay planting 3 weeks apart to stretch the harvest period.

SEED GERMINATION

Melon seeds and seedlings rot easily in cool, wet soil. Wait until the soil temperature is a least 65°F before direct seeding or transplanting. The effect of soil temperature on the speed of germination shows graphically in these test results:

Soil Temp.	59°F.	68°F.	77°F.
Days to Germ.	No Germ. (too cool)	8 days	4 days

Soil Temp.	86°F.	95°F.	
Days to Germ.	3 days	No Germ. (too hot)	

DIRECT SEEDING

Most melons are seeded direct in the garden because they sprout rapidly and strongly in warm soil. Melons are often planted in groups and the young vines trained out like the points on a star. Scoop a shallow hole the size of a dinner plate and space 4 or 5 seeds around the perimeter. Seeds can be poked into loose soil with your finger or covered with ½″ of sand, compost or other porous material. Allow 48″ between groups of plants and place the rows 72″ apart. Early varieties usually have shorter, sparser vines than midseason or late melons, and groups can be spaced 18″ apart, with rows 48″ apart.

In hot, dry soil, good germination can be achieved by making a deep furrow and flooding it 2 or 3 times with water. Drop seeds 6″ apart in the furrow and cover with dry soil. Cover the furrow with a board. Remove it gradually as the seeds emerge, to keep the seedlings in semi-shade until they form a second set of leaves.

TRANSPLANTING AND THINNING

Melons are often transplanted to the garden in zones 1 and 2, to bring them to mature size while the weather is still warm. (Melons that size-up late, don't sugar-up fully in the cool days just before fall frost).

Start seeds indoors in peat pots 5 to 6 weeks prior to the indicated transplant dates. Grow plants in full sun or under strong fluorescent light to avoid stretching, and harden for a few days. Avoid drastic and sudden exposure to cold and wind during the early part of hardening. Transplant carefully to keep the root system intact. Be sure to peel off the projecting rim of the peat pot and to cover the top of the rootball with ½ inches of soil to prevent drying out of the root system.

When planted as directed on good soil, melons are not usually thinned unless 2 or more plants emerge close together. On poor or new garden soil, plants should be thinned to 2 per hill because the soil can't support a denser population.

MULCHING AND CULTIVATION

Mulch and melons go together. Plastic mulches trap heat and keep the bottoms of melons dry. Evaporation is reduced and, in the warm soil, microflora and fauna thrive while breaking down organic matter and minerals to forms that roots can absorb. Weeds are eliminated.

Black plastic is the preferred mulch in zones 3-5. Clear plastic is much used in zones 1 and 2 because gardeners are just discovering that clear plastic can trap more heat. Weeds come up under the clear plastic but can be shaded out with cardboard at the first sign of green.

In zones 6 and 7, black plastic is used for early crops of melons but makes the soil too hot for later crops. In areas of 6 & 7 where snails, slugs and earwigs plague gardens, clean culture is the rule.

If you don't mulch, then scrape off or pull out weeds religiously while plants are young. After the plants flop and start to run, weeding becomes progressively more difficult. Up until vines begin to blossom, you can lift and turn them back onto beds to keep them out of irrigation furrows or pathways.

SUPPORTING STRUCTURES

Vining melons of all kinds can be grown on vertical supports to conserve garden space. Structures must be sturdy because melon vines are heavy when loaded with ripening fruits. Cylinders of heavy-duty fencing, stiff wire cages or a well-supported trellis of strong twine are suitable. Teepees or poles are not. You can wind melon vines around them and they will slide down in a heap unless secured by string or tape.

Developing fruits require individual support when they enlarge beyond softball size. Suspend each melon in a sling made of netting or a wide piece of cloth fastened securely at each end to make a hammock. Melons held off the ground are cleaner and usually ripen more uniformly.

WATERING

Melon vines need a steady, moderate supply of water during the early part of the season but requirements go up during hot, dry, windy weather. The numerous broad leaves present a large surface area for moisture loss. Drip irrigation has been demonstrated to increase yields of melons significantly and to bring in crops earlier. Another good way to water is with a "water wand", a long, rigid pipe that attaches to a hose. It has baffles on the nozzle to break the force of the water and make it bubble gently around the plant. With the wand you can reach to the crown of the plant where the roots are concentrated. Turn the flow to low and let the water soak in for 30 to 60 minutes at each point. If you furrow-irrigate, apply 2 to 3 inches of water weekly if no rain falls, to wet the soil to a depth of 10 to 15 inches. Neglect of regular watering after the first blooms appear can cause a reduction in yield.

CONTAINER GROWING

Standard vine melons ramble too much to make practical container plants. However, the new short vine varieties developed especially for small gardens adapt beautifully to growing in containers of 10 gallons capacity or larger. You can expect 2 to 3 full sized melons from 2 vines in a 10 gallon tub, more from larger containers. You can train the vines so that fruits are set on the surface of the soil in the container. Then, if you have to move the container, the melons or vines won't be damaged.

Set the container on concrete, flagstone or rock to gain from the absorbed and reflected heat. Raise the container on blocks to insure good drainage. Let the vines sprawl; no supports are needed.

HARVEST

Cantaloupe or Muskmelon — Watch the netting on the rind: as maturity approaches, it begins to stand out prominently against the background. Press the blossom end of the melon with your thumb; when it is no longer hard and firm but slightly springy, the fruit is nearing maturity. When perfectly ripe, the fruit stem will "slip", separate easily from the stem with only a twist, leaving a smooth disc-shaped depression in the fruit.

Casaba — The rind changes to deep yellow and the blossom end becomes slightly springy to the touch. Check also for a pleasant aroma, usually stronger at the flower end of the fruit.

Crenshaw — wait until the skin changes to a yellowish-green color, and when you can detect a sweet, fruity aroma.

Honeydew — the rind of most varieties changes to an ivory-green color, with a hint of yellow in some. The Honeydews can be picked early for ripening in storage, to avoid loss of nearly-mature fruit to rotting. But, the flavor of vine-ripened fruit is superior.

Persian — The initial indicator of ripeness is a change in rind color to light gray. About the same time, the sparse netting begins to stand out against the background. When perfectly ripe, the blossom end is slightly springy when pressed, and the melon develops a fruity perfume. Persian melons can ripen before the stem begins to slip. You will often see them sold with a piece of stem attached.

ENVIRONMENTAL PROBLEMS

Some of the problems in home production of melons come from choosing unadapted varieties. Most State Experiment Stations run trials on new varieties and evaluate earliness, flavor, production and disease resistance, sometimes in branch stations in various climate zones within the state.

Trials may at times give the nod to a variety that has high disease resistance and production but only average flavor, usually a variety bred for shipping. You have to balance their recommendations with catalog descriptions. If you ignore their recommendations, you could end up with a highly touted variety that will taste like mush when grown under your conditions. Whenever you request a list of recommended varieties from your local Cooperative Extension Service ask your Farm Advisor (County Agent) if he or she is aware of any recently introduced varieties that taste good. At least two years of testing is required for a new variety to get the blessing of the Extension Service and another year or two for it to appear on the recommended list. A dedicated Farm Advisor is always willing to help an innovative home gardener shortcut through the red tape.

Generally speaking, you should choose the latest maturing variety that will ripen during warm weather in your area. The reason for this recommendation is that the varieties that are famed for their eating quality are almost all in the medium to late-maturing classes. This does not mean that if you are in an extremely short-season area that you should try to grow these later varieties.

Failure to set fruits or failure to set a heavy crop is usually due to lack of insect pollination. There is little you can do to attract bees if most of them have been killed by insecticides but you can try interspersing plants of basil or borage, bee attracting herbs, among your crops.

Failure to set fruit may also be due to prolonged cloudy weather causing blooms to open only partially or not at all. Whatever the reason, if you have difficulty in getting good fruit set, you may have to transfer pollen from male to female blossoms with a soft brush.

Misshapen fruit can be the result of either incomplete pollination, insect damage or a combination of inadequate soil moisture and nutrient deficiencies in the soil.

Bland tasting melons may be due to a combination of cool weather and too much moisture. It has nothing to do with the form or amount of plant nutrients supplied. The best tasting melons mature during long dry spells with adequate heat to make them sugar-up fully and quickly. End of season melons are never as good as those harvested in late summer.

Fruit rot commencing at the bottom of melons can be reduced by growing the vines on plastic mulch or by placing a board or inverted tin can under each fruit.

STORAGE PRESERVATION
Melons can be stored at 40°F in a moisture proof bag for up to 1 week. They are subject to chilling injury at lower temperatures.

STORING LEFTOVER SEEDS
Under favorable conditions, at least 50% of the seeds can be expected to germinate after 3 to 5 years. Melon seeds are notably long-lived.

MUSTARD GREENS

Mustard Greens Or Leaf Mustard

Brassica juncea Var. *crispifolia*

Cruciferae (Mustard Family)

Cool Weather Annual; Seedlings Are Quite Frost Hardy, Mature Plants Are Moderately So

Killed By Prolonged Freezing Weather

Full Sun Or In Late Summer, Light Shade

Ease Of Growth Rating -- 1

DESCRIPTION
A drive across the South in late fall will reveal patch after patch of mustard greens, recognizable by their light green foliage, usually side-by-side with a dark green patch of turnips grown more for greens than for roots. The mild, open winters in zones 6 & 7 keep greens flourishing until a "Blue Norther" blows through in January, freezing the mustard but usually sparing the hardier turnips.

Northerners, too, can grow mustard greens, and should. Mustard produces large quantities of nutritious food in a short time on a small area. The plants grow 12 to 18 inches in height and, when well-grown, their individual leaves may reach 9 to 12 inches in length. However, the smaller inner leaves are more tender, less peppery and can be eaten midrib and all. The upright plants are brittle and are usually planted in sufficient quantities for harvesting whole. Individual leaves can be taken, leaving the plant to produce more, but this is time consuming.

Many cultivars of mustard greens are gown in the Orient . . . types never seen here because they were developed for semi-tropical conditions. Generally the Japanese types are more pungent than those developed in China and Taiwan, partly because the peppery flavor holds up when the greens are pickled.

Mustard greens originated in Europe but are not grown extensively there now. It is their loss; consider-

ing the multiple uses of mustard greens. They are more than a homely potherb for boiling with side meat; the tender, frilly young leaves make an excellent salad ingredient or they can be served as "wilted greens" with piping hot bacon drippings poured over them.

SITE

Spring-seeded mustard greens grow faster if they are planted on well-drained soil in full sun. Plantings made from midsummer on can be shaded. The warmer the climate, the more shade the plants can stand. An ideal arrangement is to have the shade coming from deciduous trees south and west of the mustard patch. The trees drop their leaves when the days grow short and when the greens need exposure to full sun.

SOIL

Mustard greens will grow well on most garden soils; you can even grow passable greens on poor soils but they will be pale green and rather stringy. Mustard greens tolerate a wide range of soil pH but grow best at levels of 5.5 to 6.8. This crop is a heavy feeder and requires a good supply of nitrogen to produce maximum yields of leaves that are tender at large sizes.

SOIL PREPARATION

Mustard greens are so robust that they are often broadcast on roughly worked soil. If the greens are harvested during the rainy season, experienced gardeners lay old boards down the pathways so they can get in to harvest greens despite soggy soil. Another arrangement is to "drill" the seeds close together in straight, single ridged-up rows for improved drainage.

In all except zones 1 & 2 mustard greens can be sown in midwinter; the seeds will not wash away but will be gradually covered by the roughly worked soil as the snow melts down. The seeds will not rot; they will come up at their appointed time in the spring.

PLANTS PER PERSON/YIELD

Grow 12 to 15 plants per person for fresh use only. This should yield 6 to 8 pounds of greens if harvested at full maturity.

SEED GERMINATION

Mustard seeds are strong growing at a wide range of soil temperatures and should sprout within four to ten days. The usual germination of fresh seed is 70 to 85 percent.

PLANTING DATES BY CLIMATIC ZONES

Mustard greens are too strong to eat during hot weather and seeds come up poorly in hot soil.

Therefore, spring seedings are made very early to mature before hot weather, and later plantings are delayed until the soil begins to cool. In zones 1 & 2 hot soil isn't a problem. Crops for fall harvest are direct-seeded after the days begin to grow shorter, in order to avoid bolting.

DIRECTIONS FOR DIRECT SEEDING

Mustard greens are difficult to transplant, even at small sizes, and are almost always direct seeded. Sow about a dozen seeds per foot of row and cover to a depth of ¼ inch, or ½ inch in hot soils in late summer. Wide band planting is often used in fertile soil in older gardens with few weed problems. A typical wide band planting is made to the width of a garden rake, with seeds spaced on one-inch centers. Imbed the rake deeply and pull it through the soil to leave furrows behind the tines. Scatter the seeds thinly over the band and they will tumble into the furrows. Close planting will give you at least twice the optimum population of plants and will produce plenty of thinnings to pull and eat when about 8 inch in height.

DIRECTIONS FOR TRANSPLANTING AND THINNING

Mustard is tap-rooted and difficult to transplant. Seeds sprout and grow so fast in the garden that transplanting is counterproductive.

Begin thinning when the largest plants reach 6 to 8 inches in height. Repeat every three or four days to keep the plants from crowding each other and growing tall and spindly. Ideally, your plants should be standing 6 to 12 inches apart at maturity. If your first thinnings don't give you enough for a "mess" of cooked greens, use the tender young leaves in salads.

Mustard plants need no supports. They can be knocked over by high winds and rains but the supple stems will turn upwards and resume growing.

MULCHING AND CULTIVATION

Mulching isn't warranted for spring planted mustard greens, but will pay off for your fall crop. Apply an organic mulch to moderate depth as soon as the plants have 6 to 8 leaves. Don't mulch winter greens in zones 6 & 7; the organic matter will keep the soil wet and cold and will provide a hiding place for snails, slugs, earwigs and sowbugs.

Large mustard patches on fairly weed-free soil in zones 6 & 7 are rarely mulched but are broadcast so thickly that the greens shade out most of the weeds. Spring-planted mustard is seeded so early that it requires little cultivation for weeds. Mustard for fall harvest usually follows another crop. Customarily, the soil is worked up and watered heavily to sprout weeds seeds; these are hoed out before the mustard seeds are planted.

WATERING

Large patches of mustard greens are often left to the vagaries of nature, and complete failures are frequent during droughty seasons. It would seem more sensible to grow the mustard nearer the house, where a sprinkler could be set to run for about an hour before being moved to soak another area. Set a water glass near the sprinkler to get a rough measure of the amount of water you are applying. Two to three inches should wet the soil to a depth of about 12 inches, deeper in sandy soil. Mustard greens use lots of water, especially when grown in late summer when the leaves are transpiring heavily.

CONTAINER GROWING

Mustard greens grow well in containers and are often produced as a spring crop, planted in containers very early indoors and, after harvest, replaced with a summer crop such as peppers or tomatoes. Mustard plants can grow surprisingly tall in the porous planter mixes and containers should be of at least 3 to 5 gallon capacity to support them.

Late in the growing season, when the summer container vegetables begin to decline, sow mustard seeds around them for a cold-tolerant fall crop that can be moved under cover to prolong the harvest.

ENVIRONMENTAL PROBLEMS

The problem most often reported with mustard greens is "bolting", the shooting up of flowers and seedheads. The only way to avoid bolting is to plant early in the spring and again after midsummer. When long, warm days come, nature has programmed mustard to go to seed in order to reproduce itself.

Some gardens occasionally produce tough plants of mustard greens. These result from infertile or dry soil which will not permit quick development of tender, succulent leaves. If drought causes mustard greens to become tough, discard the stems and the central midrib, and use only the leaf blades.

STORAGE

Mustard Greens can be kept in the refrigerator 5 days in a moisture-vapor proof bag.

STORING LEFTOVER SEEDS

When placed in a sealed container with a desiccant and stored in a refrigerator, mustard seeds should germinate 50 to 75 percent after three years.

OKRA

Okra (Gumbo, Lady's Fingers)

Abelmoschus esculentus* Formerly *Hibiscus esculentus

Malvaceae (Mallow Family)

Warm Season, Heat Resistant Annual

Damaged By Prolonged Cool Weather, Killed By Light Frost

Full Sun

Ease Of Growth Rating, South -- 2; North -- 3

DESCRIPTION

Okra is a large, robust, tropical-looking plant, 3 to 8 feet tall and 2 to 3 feet wide, and is grown for its pods which follow attractive yellow flowers in the axils of leaves.

While the authors of gourmet cookbooks were making fun of okra, or ignoring it, this slippery vegetable was invading the North. Their oversight was due to a basic misunderstanding about the "roots" of okra. Northern authors associated it with Creole or Cajun cooking, which is largely restricted to southern Louisiana, southern Mississippi and parts of east Texas. What they did not know was that okra came to all the southern colonies with slaves or traders who brought the seeds directly from Africa or from the West Indies. It, together with another African vegetable, southern peas, quickly spread across all the South because these two vegetables were among the few that would produce food throughout long, hot summers.

The major use of okra pods is not, as cookbooks would lead you to believe, in gumbo. Instead, it is fried, boiled lightly with bacon drippings or butter, added to soup stocks and pickled in brine. Years ago, poor families dried the pods and stored them for winter use of their nutritious seeds. This practice was copied from slave families whose ancestors had dried okra pods in Africa.

Tremendous progress is being made in okra breeding. Only a few years ago, typical okra cultivars were spiny and grew extremely tall in the South, sometimes to 12 feet or more. Garden columns were replete with photos of gardeners standing atop ladders to harvest okra. Spineless varieties were developed first, then the process of dwarfing began. Now, breeders are concentrating on F_1 hybrids that are not only early bearing, spineless and dwarf, but that also have multiple stems or short internodes between leaves. Okra pods are borne in the axils of leaves, so the more leaf joints, the more potential for increased production.

Perhaps the greatest force in making okra a national vegetable was the migration, during and after WWII, of many Southerners, blacks and whites, to northern States, and the reverse migration to Sunbelt states, now underway. Frozen food producers followed these changing demographic patterns and

OKRA

food tastes with packaged okra, and it began appearing in supermarkets all over the USA.

SITE

Leave plenty of space in a warm, sunny area for okra to spread out. If your yard has a cold, windy micro-climate, put your okra bed where it will be protected by a wall or a fence and where reflected heat can bounce back into the plants and make them grow more rapidly.

SOIL

It is difficult to grow okra on fine, sandy soil because of its need for a steady, adequate, but not excessive supply of soil moisture and plant nutrients. If you have dry, sandy soil, mix in moderate amounts of organic matter to produce fine particles of humus which will supplement the missing clay particles, and mulch the crop. Okra is only mildly tolerant of acid or alkaline soils and does best at pH levels of 6.0 to 6.9. Okra will not fruit well if excessive nitrogen is present in the soil. If your soil drains slowly, ridge it up for okra rows.

SOIL PREPARATION

Okra is tap-rooted and likes a deep seedbed. Only moderate amounts of organic matter should be incorporated in order to avoid excessive release of nitrogen when warm soil accelerates its breakdown.

Ridged rows or raised beds keep heavy soils drier and warmer but should not be used for lightweight soils which would dry out excessively. Well-drained soil is less likely to foster the soil-borne diseases which occasionally attack okra.

PLANTS PER PERSON/YIELD

In the North, grow 10 to 12 plants for each person; in the South (where okra is more productive) 6 to 8 plants. In the South you should harvest 12 to 15 lbs. of baby okra from this planting. Bear in mind that 6 to 8 plants will bush out to take up 12 to 16 feet of row; okra is not space efficient.

SEED GERMINATION

Okra seeds are customarily soaked overnight in tepid water (often in a thermos to keep the water warm), dried and planted promptly. With this treatment, germination should be faster than the laboratory range:

Soil Temp.	50°F.	59°F.	68°F.	77°F.
Days to Germ.	No Germ.	27 days	17 days	12 days

Soil Temp.	86°F.	95°F.	104°F.
Days to Germ.	7 days	6 days	6 days

Planting at the upper range of temperatures should be avoided, if possible, because hot, drying winds can kill seedlings soon after emergence.

DIRECT SEEDING

Garden soil temperatures seldom if ever rise to the optimum for okra seed germination in zones 1 and 2 and, during cool summers, in zone 3. For this reason, direct seeding is not recommended for zones 1 and 2, nor for zone 3 except with the condition that transplanting in peat pots works best in most years.

Delay seed planting for two to three weeks after frost danger has passed. Dry seeds and sow in single rows, placing seeds 6 inches apart. Or, plant two seeds per hill, with 12 inches between hills. Cover seeds ½ to ¾ inches deep. In heavy clay soils, cover seeds with sand to avoid crusting after rains.

TRANSPLANTING AND THINNING

Okra is tap-rooted and transplanting always results in some setback for plants. Yet, it must be done in zones 1, 2 and sometimes 3 in order to make the plants bear before cool fall weather sets in. Start seeds indoors in peat pots no more than a week or two prior to the frost free date. Set the pots in the garden as soon as two to four leaves have formed and cover the seedlings with bottomless plastic jugs that act as greenhouses. Space the seedlings one ft. apart and leave plenty of room for them to bush out. The plants will grow tall and shade adjacent rows of vegetables if you don't plan ahead.

Thin okra to stand 1' apart in fertile soil and 18" apart in poor or dry soil.

MULCHING AND CULTIVATION

Okra is often mulched with black plastic in zones 3 through 5 to build up heat in the soil while maintaining a uniform level of soil moisture. It is usually laid over ridged rows or raised beds and the edges battened down with soil. Clear plastic would work better in zones 1 and 2 but weeds would need to be shaded out after the soil has warmed. Black plastic can raise the soil temperature too high in zones 6 and 7; organic mulches applied after the crop has grown knee-high work better.

Summer weeds can be a problem until okra outgrows them. Scrape off weeds with a hoe. Nothing is gained by hilling soil up around okra plants.

SUPPORTS AND STRUCTURES

Okra plants are stiff and woody and usually don't need supports. They can be knocked over in thunderstorms but can be righted and supported with a stake until new anchor roots strike.

WATERING

Okra is deep rooted and, once well established, will thrive if it receives a good rain or irrigation every two to three weeks. Soak the soil to a depth of 2 feet by applying 4 to 6 inches of water. This much water is difficult for clay soils to soak up without runoff and is best applied in two stages.

CONTAINER GROWING

Okra will grow in deep containers of heavy soil but usually isn't, except as a decorative curiosity. The decorative variety Red River is excellent for this purpose. Grow two or three plants per 30-gallon container. Set the container in a corner so it won't blow over. So many plants of okra are needed to produce a meal at a picking that container production is insignificant.

ENVIRONMENTAL PROBLEMS

Okra has few problems except for its habit of growing to a size partly determined by the warmth of the climate and length of season. Many northern catalogs will give their local experience in the size and height of plants. Dwarf cultivars may be listed at 2½ to 3 feet in height and standard cultivars as 4 to 4½ feet. However, where summers are long and warm, the plants can grow considerably larger and catch the gardener by surprise if he or she has not provided ample space.

Another problem is the cessation of bearing when the gardener has not picked frequently and has permitted old pods to hang on the stalks. Pods should be picked every 2 to 3 days to encourage the formation of new sets of okra. Late in the season, the pods on the top of the plant can be permitted to remain and to dry; they make curious and novel additions to dried flower arrangements.

Okra plants are so sturdy that they will weather much damage and continue producing. They won't do well where Southern Root Knot nematodes are entrenched. You may wish to grow and plow under a crop of the special French marigolds that reportedly repel nematodes for up to two years.

HARVESTING

Every 2 to 3 days, snip off the young pods. If a few escape and become woody, clip and discard them. Some cultivars, especially during cool spells, will produce pods that are tender to a length of 4 to 6 inches but most are harvested at 3 to 4 inches.

STORING LEFTOVER SEEDS

Seeds should keep for 3 years and still germinate 50 percent. Store them in a sealed container along with desiccant. Put the jar in the refrigerator during summer months. The combination of high heat and high humidity during southern summers is very destructive to seed germination and longevity.

ONIONS AND SHALLOTS

Onions and Shallots

Allium cepa (Common Onion, Dry Storage Onion)

Cool Season Biennial, Grown As An Annual

Winter Hardy In Most Climates

Ease Of Growth Rating, From Sets And Seedlings -- 2; From Direct Seeding -- 4; From Transplants You Grow Yourself -- 5

Allium cepa Proliferum Group (Egyptian, Top, Tree or Walking Onion)

Cool Season Perennial, Winter Hardy In Most Climates

Ease Of Growth, From Divisions Or Bulbils -- 1

Allium cepa Aggregatum Group (Shallots Or French Shallots, Potato Or Multiplying Onions)

Cool Season Perennial, Winter Hardy In Most Climates

Ease Of Growth, From Cloves (Bulb Divisions) -- 2

Allium fistulosum (Welsh Onion, Japanese Bunching Onion, Spanish Onion)

Cool Season Perennial, Winter Hardy In Most Climates

Ease Of Growth Rating, From Direct Seeding -- 3

FOREWORD

The saga of the onion family reads like a Russian novel, full of obscure relationships and, for its members, double and triple nicknames and highly individual personalities. That such a complex family developed over the centuries of onion/man interdependence is not surprising, for onions have been a key food crop to many civilizations. The onion family has grown and evolved at the hand of man; untrained farmers as well as plant scientists have left their mark.

This condensation of the onion family saga may help you not only to select just the right onion family members for your garden, but also to understand why critical selection of kind and variety is so important to good onion performance.

To familiarize you with the onion family rapidly, we have broken it into four clans. Each clan looks, tastes and is grown differently: Common Onion, Egyptian Onion, Shallots and Welsh Onion. We will deal with the common onion in considerable detail because it is the most popular and complex of the onion clans.

DESCRIPTION
COMMON ONION

The common onion, despite its many colors, sizes and shapes, goes through a singular process of development. You plant an onion seed and, in one season, it grows into a scallion, then matures into a bulb. The bulb is usually harvested and stored for later use. Should the bulb be left in the ground, or replanted the next spring to resume growth, it will shoot up a flowering stalk and set a cluster of seeds. Thus, in two years, its biennial life cycle is completed.

DAYLENGTH FACTOR. One of the least publicized peculiarities of the common onion clan is its so-called "day-length response." More accurately, it is a "night-length response." When you buy onion seeds, seedlings or sets locally, you will get varieties whose biological clock is set for your area. But, when you order from a catalog published by a regional company in a latitude greatly different from yours, you may unwittingly specify a variety that will not bulb in your locality.

Varieties of common onions can be classified as to the length of day required to induce bulbing. "Long day" varieties require longer days to form bulbs than do the "Intermediate" or "short day" varieties.

During the growing season, the day-length increases as one moves further north. In the northern states and Canada, at and beyond 40° North, long day varieties must be used to produce medium and large size onion bulbs for storage and cooking.

However, remember that many northern areas of the world have longer summer days than prevail, say, in Maine or Southern Ontario. Such areas use what are called "extra long day" varieties suited to their very high latitudes. If you import seeds adapted to the latitudes of northern Europe and plant them in the northern USA or southern Canada or southward, they will not receive a sufficiently long day to induce the development of good sized, fully mature bulbs for long storage.

Long day varieties adapted to the USA and southern Canada contain the only types suitable for long term bulb storage. These varieties are called "hard storage onions." Short day onions are too soft and too high in moisture to store for more than a few weeks.

Short day onion varieties are preferred for planting zones 6 and 7. When planted in late summer or early fall they initiate bulbing during shortening days and grow on to mature bulbs in the spring.

If short day varieties are spring planted in the north, they will receive the full length of day required to induce bulbing early in their life cycle while the plant is still small, and the resulting bulbs will be correspondingly small. On the other hand, if you plant long day northern varieties in the south in the spring, they will never receive long enough days to bulb and will grow vegetatively with a swelling at the base, not producing anything typical of a good bulbing onion.

You can manipulate the day-length factor to your advantage if you wish to produce pickling or pearl onions up north. These are special short day varieties and when they are spring planted in the north will mature as very small bulbs for pickling or canning.

HARD VS. SOFT ONIONS. The large, tender slicing onions are universally loved for their mild taste. But, they will store for only short periods as compared to hard bulb varieties which have many layers of dry scales and no "necks". Very hard bulb storage onions have a strong flavor, too pungent for all but the brave to eat as thick, raw slices.

DESCRIPTION
EGYPTIAN ONIONS

"Starts" of these Egyptian onions have been handed down in families for generations and passed around to friends. Now, a few seed companies offer them. The mature plants split and form a ring of sturdy shoots around the mother plant. These grow into tall, round blades, topped by clusters of bulbils or bulblets which can have gray-green, reddish or purple skin. In the late summer bulbils can be pulled off and planted to grow through the winter to produce early spring scallions.

The tops of Egyptian onions are a bit tough but can be chopped for frying or for use in soups or stews. The taste may be a mite strong for gardeners who are accustomed to delicate scallions from bunching or common onion varieties. Yet, for a standby onion to fill in before annual onion crops are ready and after the winter supply is exhausted, nothing beats a few plants of Egyptian onions.

Eygptian onions have erroneously been called "Rocambole", which is a garlic with coiled leaves.

SHALLOTS, POTATO AND MULTIPLIER ONIONS

To shallot aficionados, grouping them with potato or multiplier onions might seem to be mingling the sublime with the ridiculous. But genes, not man, decide the distinction. All three are winter hardy perennials and have large bulbs which divide into a number of cloves somewhat resembling garlic. Some varieties set seeds, but sparsely, and some produce topsets or bulbils.

Shallots produce bulbs that split into distinct sections called cloves, loosely connected at the bottom. When you plant shallot bulbs, they shoot up a dense cluster of 6 to 12 scallions. (If you think that shallot cloves are the epitome of flavor, just try the scallions!) Save some of the scallion clumps for dividing and replanting in late summer or early fall to replenish your shallots with home-grown bulbs the following summer. You can also force shallot cloves indoors, to produce scallions.

ONIONS

The potato onion looks something like shallots but the leaves are broader and the mother bulbs, at first glance, appear to be large, single bulbs. You have to peel off skin to expose the numerous cloves which are flattened above and below into a broad "potato" shape. Potato onions have sparse flower clusters and set a few seeds but are usually propagated by dividing bulbs into sets and planting them in the late summer or early fall.

Potato onions have been cultivated in England since 1796. Multiplier onions are often confused with potato onions. They look and divide very much the same, except that the individual cloves are shaped more like garlic cloves and less like little potatoes. Multiplier onions have undergone a great amount of differentiation in the process of being handed down and selected by generations of home gardeners.

WELSH, JAPANESE BUNCHING, OR SPANISH ONIONS

This onion clan has no more connection with Wales, Japan or Spain than Egyptian onions have with Eygpt. In recent years, Japanese plant breeders have concentrated on improving varieties and, in the garden trade, the term "Japanese Bunching Onion" is often used. The inscrutable United States Department of Agriculture prefers the term "Welsh Onion" for the Japanese variety formerly known as "Prolific White Bunching Onion".

The Welsh onion is the one you most often see in grocery stores as bunched scallions. No matter how long you let the plants grow, they will not produce large bulbs. Instead, Welsh onions produce clumps of winter-hardy scallions which resemble overgrown chives. Some may have a small swelling at the bottom of the stem, but no bulb. Welsh onions can be reproduced from seeds or by dividing the long lived clumps. Some varieties of these bunching onions grow single stems from each seed; others grow from 3 to 5 plants from a single seed, obviously a good buy.

SITE

In zones 1 through 3, selecting the site for common onions is more important than elsewhere. Bulbing onions are a warm weather crop and to succeed in short season areas they need full sun, and preferably a south tilted slope. Shallots, potato onions, Egyptian and Welsh onions don't form large bulbs and are not so critical as to site.

Plant onions where they can be protected from foot traffic. They have a dense, compact root system concentrated near the surface of the soil, and soil compaction can retard development.

SOIL

Common onions are particular about soil and prefer sandy, sandy loam or organic muck soils, especially in zones 1 and 2. Further south and in the west, onions will also grow on well drained clay loam heavily modified with well decomposed organic matter, but they will not grow on tough, hard clay. On the heavier soils, raised beds will pay for themselves in increased yields.

Common onions are only slightly tolerant of acid soil and grow best in a pH range of 6.0 to 6.8. They remove little nitrogen from the soil, a moderate amount of potash and a significant amount of phosphorus.

Egyptian onions, shallots and Japanese bunching onions are not as particular about soil as common onions but, nevertheless, prefer fine textured, sandy, sandy loam or muck soils.

In areas of moderate rainfall, 25 to 30 inches annually, and where the soil has been improved for some time with manure and green manure crops, a single application of preplant fertilizer can support an onion crop from planting to bulbing. Work toward that goal by incorporating all the organic matter you can beg, borrow, buy or grow. Be sure to work it in about a month before planting to have the decomposition well started before planting. Apply lime to acid soils the prior year; if you wait until just before planting use hydrated lime instead of limestone, but handle it carefully; it is caustic.

On heavy soils, build up beds to promote drainage and to keep rainwater from standing around the maturing bulbs of common onions.

Commercial growers in the north grow green manure crops and spread manure over them in the fall for turning under in the spring about a month before seeding onions. In the south, where onions are direct-seeded in the late summer or fall, summer annual green manure crops such as cowpeas or soybeans are used.

PLANTS PER PERSON/YIELD

For green onions or scallions, grow 30 to 40 plants for each person. Split the crop into two plantings, made two to three weeks apart to spread out the harvest.

For bulbs, provide 25 plants per person if you plan to eat the bulbs fresh. For storage bulbs, plant 50 to 100 per person if you have room in a cool, moist storage area for such a large crop. At ¼ pound average per bulb, this is a lot of onions . . . up to 25 pounds per person. Southern and western gardeners, note: short day varieties do not keep well.

Egyptian onions, 6 to 12 plants will suffice. Shallots or potato onions, plant all the space you can spare; well grown shallots are delicacies and will make much appreciated gifts.

SEED GERMINATION

Common and Welsh onions can be grown from seeds. Onion seeds are notoriously short lived and are not especially vigorous, particularly when kept for more than a year.

Soil Temp.	32°F.	41°F.	50°F.	68°F.
Days to Germ.	135 days	31 days	13 days	7 days
Soil Temp.	77°F.	86°F.	95°F.	104°F.
Days to Germ.	5 days	4 days	12 days	No Germ.

DIRECT SEEDING

Seeds are the least expensive method for producing common onions and the usual way to get a start of Welsh or Japanese bunching onions. Direct seeding is usually preferred for fertile, well drained soil, providing you can get an early start. Onion bulbs grown from direct seeding are always of the highest quality.

Space seeds about ½ inch apart and cover to a depth of ¼ to ½ inch. If you plant only a short row, cover with sand, vermiculite or compost, otherwise, use garden soil.

Onion seedlings look so much like grass that straight row seeding is preferred to broadcasting or planting in wide bands.

PLANTING SEEDLINGS AND DIVISIONS

Onion seedlings grow slowly and need to be cultivated and weeded carefully for several weeks. For this reason, some gardeners prefer to plant seeds early indoors and to transplant seedlings to get an early start on the season. Of the three ways to produce common onions — seeds, seedlings and sets, seeds offer the greatest choice of varieties and the best likelihood of producing large, uniform bulbs.

Plants of onions are also known as seedlings or transplants, these resemble scallions. You can buy a limited selection of varieties that have been grown to pencil size by commercial specialists, but these will be of the soft, non storage varieties.

In the north, seedlings can be grown indoors for transplanting at pencil size or slightly smaller. Southern gardeners enjoy a distinct advantage because, with short day autumn planted varieties, they can start seeds out of doors in protected nursery beds for transplanting.

To get big seedlings from indoor plantings, start 8 to 10 weeks prior to the recommended transplanting date for your zone. Use flats, pots or other containers and sow seeds thickly, about ⅛ inch apart. Cover to a depth of ¼ inch. Sprout and grow seedlings under fluorescent lights with the tubes only about 2 inches above the plants. It is exceedingly difficult to grow good, big seedlings using only the weak sunlight of late winter days. Under lights, the seedlings will expand and grow close together, but no problem; you can pull them apart at transplanting time. Harden for several days before transplanting. Set the seedlings in the soil to the same depth as they grew in the container. Seedlings should be large and vigorous to "take" strongly in the garden. Small, spindly seedlings are virtually useless.

Transplant seedlings 1 inch apart for scallions, 3 to 4 inches apart for large bulb production. Make rows 24 inches apart to let you pull soil up around scallions.

STARTING FROM SETS

Onion sets are more expensive than seeds or seedlings, but are so convenient that gardeners with small plots often use them to produce scallions sooner than from direct seeding and to produce a few early bulbs for storing. Sets are usually sold by the pound and, in retail stores, sans variety name. You buy them by red, yellow or white colors.

Sets bear explanation. They are an example of how man can manipulate the biennial cycle of onions to his advantage. Seeds of certain varieties, if planted thickly and not thinned, will form miniature bulbs called sets. These small sets, ⅜ to ¾ inches in diameter, are not midget adults, they are juveniles, as yet incapable of the weighty process of reproduction. When stored over winter for spring planting, sets will resume vegetative growth and form large scallions, then bulbs.

Sets would seem to be *the* way to grow onions — easy to plant, fast growing, sure to bulb. They are all of that, but they are more expensive than seeds or seedlings, are available in only a few varieties and are usually sold only during a short period of spring in retail stores and by mail order.

Purchase sets in sizes ⅜ to ¾ inches in diameter. Larger sets will bolt to seeds and smaller ones will come along slowly. Examine the sets and you will see the root disc on one end of the bulb. Plant this end down. Place the sets upright and press them firmly into 2 inch deep furrows. Space sets 2 inches apart. Cover the sets with loose soil. If for scallions only, double or triple rows can be spaced as close together as 4 inches but you will need to allow at least 24 inches between each set of rows to provide enough soil to pull up around the scallions for blanching long stems. One pound of sets (approximately 200) will plant a single row about 40 feet long.

If planted to produce hard storage bulbs, sets should be placed no closer together than 4 inches, in double rows 6 inches apart. You can kill two birds with one stone by doubling the planting rate and harvesting every other plant early for scallions.

STARTING FROM CLOVES OR DIVISIONS

Shallots and potato onions are usually purchased as bulbs in early spring in zones 1 through 5 and in the fall in zones 6 and 7. The bulbs can be separated into cloves and planted 3 to 6 inches deep in the north and 1 to 2 inches deep in the south. Shallow planting in the north could lead to loss of plants during the winter due to heaving by frost.

Plant shallot or potato onion cloves 6 to 8 inches apart, root disc down, in rows 12 to 15 inches apart. Because shallots, potato and multiplying onions are winter hardy, mature clumps can be left in the ground over winter, except where winters are quite severe and snow cover is sparse. These will produce green onions similar to scallions early the following spring.

Shallots may also be planted in pots in the fall and grown indoors by a cool, sunny window where they will produce green tops for use throughout the winter.

MULCHING AND CULTIVATION

Onions are not ordinarily mulched but, in sandy soil that tends toward dryness, a light to medium organic mulch will conserve moisture. It should not be applied until the soil is thoroughly warm, at least 70°F. Don't walk on the mulch near onions or you may crush the dense mat of feeder roots near the surface of the soil.

Single onion rows or sets of rows are usually planted far enough apart so that soil can be scraped up from the row middlings and drawn up to the plants as they grow. In doing this you should be careful not to push over the young seedlings. Pulling up soil should be done early in the season so that you do not scrape off the extensive mat of roots which develops as bulbs begin to size up. Scraping the soil also serves to cultivate and remove most weeds.

Despite instructions in some books, don't pull the soil away from the roots which are to remain for bulbs. Leave it over them; some will wash away due to rain or sprinkler irrigation; the overburden of soil will not bother them.

FERTILIZING

Onions are heavy feeders (inefficient feeders would be more accurate, but the net result is the same — you need to apply a large amount of plant food in order to get good growth of onions). One or two light sidedressings of poultry manure during the season will bring bulbs to full size. In fact, some of the most beautiful onions ever produced were grown with sidedressings of manure.

WATERING

Onions are a shallow rooted crop and their growth can be checked by allowing them to go dry. If no rain falls for 7 to 10 days, irrigate with 1 to 2 inches of water to wet the soil to a depth of 5 inches. In arid western climates, every third or fourth irrigation should be a deep one to drive salts to lower levels in the soil.

CONTAINER GROWING

You can produce bulbing onions in containers but, considering the long period of time from seeds or seedlings to mature bulbs, they are probably not an efficient, economical crop. Scallions make a much more practical crop.

Scallions are an excellent crop for containers, and will grow best in shallow, rather wide containers. You can also stick sets around the bases of larger, later maturing plants.

Welsh or Japanese bunching onions growing in the garden can be divided in late fall and potted for indoor growth. They will grow on a windowsill until the days become too short and gloomy.

PROBLEMS

The most common complaint among onion growers, is, "Why can't I grow as big onions as my neighbor?" One secret to growing large, tender onions is to start seeds of soft-bulb varieties very early indoors to get large seedlings for setting in the garden as soon as the soil can be worked. You gain up to 3 or 4 weeks over direct-seeded onions. Your transplants should be nearly as large as scallions you buy from the store. Some mail order catalogs offer well-grown seedlings of these varieties and, if you customarily grow large quantities of these onions, you might consider purchasing your seedlings. Remember, onion sets will produce small, hard bulbs.

In zones 1 through 3 onions need special care to produce large bulbs. They are a warm weather crop and love warm, well drained soil which is well supplied with nutrients and has a pH of over 6.0. A northern gardener with heavy soil with a northern exposure has two strikes against him or her. An ideal environment for a bed of onions in a northern garden would be:

1. A southern exposure. This can make a big difference in maturity dates.
2. A light soil which drains well and heats up quickly.
3. A soil pH of over 6.0., preferably 6.5.
4. High fertility. Soil that has been well composted or manured, with fertilizer added if necessary.
5. Early spring seeding. Seeding or transplanting should be done as soon as the soil is completely thawed and can be safely worked.

6. Weed control. Frequent cultivation as onions despise shade thrown by large, aggressive weeds.

By providing these conditions, you can grow medium to large onions, as long as the crop is not affected by diseases or damaged insects.

Beginners have to learn to avoid rotting of stored onions. If temperature and humidity requirements are met, this problem can be reduced by discarding bulbs with thick necks or injuries, by improving the ventilation and by experimenting to find hard bulb varieties with superior keeping qualities. Also, discard any bulbs which show any sign of disease before or shortly after beginning the storage term. They will not keep well.

Bulbing onions grown from sets will occasionally shoot up seed stalks prematurely. This can be due to the use of sets that are too large or that have been stored at too warm temperatures by the producers. If spring planted sets shoot up summer seeds in your garden, demand a refund from the retail store; it isn't your fault. Be sure to buy sets more than 3/8 inches but less than from 3/4 inches in diameter.

Scallions are easy to grow from sets and almost as easy to produce seeds. The most common problem with the production of scallions is the failure to plant seeds sufficiently early. The newest way, and it works well, is to plant the seeds early out of doors and barely cover them with sand. Then, place a clear plastic sheet over the row and bury the edges to keep it from flying away. As soon as the onions have emerged, raise one corner of the sheet slightly to keep the air under the plastic from overheating and then, in a few days, remove it completely.

STORAGE

Green onions are highly perishable. Store in a moisture proof wrap and use within 3 days. Onions must be mature and thoroughly dry for storage. They should be stored in an open mesh bag in a dry well-ventilated area. Onions can also be braided by their stems and hung. Attics or unheated areas are ideal. Slight freezing doesn't harm the onions if they aren't handled while frozen. Store mature shallots in a mesh bag or braided in a cool dry place as for dry onions.

STORING LEFTOVER SEEDS

Do not attempt to store leftover onion seeds. They are among the shortest-lived seeds and will grow poorly on second and succeeding years. Purchase only what you need each year.

PARSLEY

PARSLEY

Parsley -- Leaf Parsley And Rooted Parsley (Hamburg Or Turnip-Rooted)

Petroselinum crispum **And** *Petroselinum crispum var. tuberosum*

Umbelliferae (Parsley Family)

Cool Season Biennial Or Short-Lived Perennial, Grown As An Annual

Killed By Prolonged Freezing Weather

Full Sun Or Partial Shade

Ease Of Growth Rating, From Transplants -- 2; From Plants You Grow From Seeds -- 5; From Direct Seeding -- 8

DESCRIPTION

One could quibble forever over whether parsley is an herb or a vegetable. We are listing it with the vegetables because it is usually grown among other vegetables in home gardens, and not singled out for a special herb garden. Leaf parsley is one of the most nutritious green vegetables and deserves to be used for more than a garnish, breath freshener or soup additive. Perhaps the answer lies partly in more extensive use of the "plain" or "single" parsley that reputedly holds its flavor better in cooked dishes than the curly types, and greater uses of parsley in salads, sandwiches and teas.

Leaf parsley grows in small mound-shaped plants, six to eight inches high and twelve to sixteen inches across. Each plant is made up of dozens of stems.

Rooted parsley is a novelty in this country but a popular vegetable in Europe where, in the cooler climate, it develops large, succulent roots 1 inch in diameter by 6 inches in length. The roots resemble down-sized parsnips, only more slender, and are creamy white.

Like parsnips, the roots of rooted parsley become sweeter with cool fall weather and can be left in the ground for extended periods if protected. Once you taste these sweet roots in soups or stews, or boiled and served with butter, you won't be happy with the kind from the produce market that are often harvested during warm weather.

The leaves of rooted parsley are edible and are used for drying and cooking. The older leaves tend to be tough, so choose the younger leaves. For salads, chop the leaves of rooted parsley finely to minimize the chewy texture.

Parsley was used as an herb before the time of Christ and has 'gone wild' in many favorable climates.

SITE

Both rooted and leaf parsley are long season crops and should be placed where they will not get in the way when you are harvesting salad vegetables. So little leaf parsley is required that the decorative plants are often placed here and there among flower beds. In zones 5 through 7, both leaf and rooted parsley benefit from light afternoon shade during the hottest part of summer. Even with shade, parsley does poorly and may die in midsummer in zones 6 & 7.

SOIL

Parsley is moderately tolerant of soil acidity and grows best at pH levels of 5.5 to 6.8. An organic matter level of 2 to 3 percent, enough to turn soil slightly dark, helps to extend the availability of nutrients and to keep a constant supply of deep green leaf parsley coming off. Rooted parsley, like other root crops, appreciates a good level of potash.

SOIL PREPARATION

Parsley is a deep rooted crop and likes lightweight, well drained soil. If you grow it in heavier soil, build up beds or plant it on ridged rows.

PLANTS PER PERSON/YIELD

Leaf parsley — Grow 2 to 3 plants for each person. This will yield several dozen sprigs at each cutting. Grow double this amount if you plan to use parsley extensively in cooking or for drying.

Rooted parsley — Grow 12 to 15 plants for each person. This will yield 8 to 12 pounds of roots plus small harvests of sprigs for use as a garnish and in cooking.

SEED GERMINATION

Soil Temp.	50°F.	59°F.	68°F.
Days to Germ.	29 days	17 days	14 days

Soil Temp.	77°F.	86°F.
Days to Germ.	13 days	12 days

Although parsley seed germinates faster in very warm soil, avoid planting during the heat of the summer because parsley plants grow best at a temperature range of 60-65 degrees F.

DIRECT SEEDING

To hasten germination, soak the seeds overnight in tepid water before sowing. Even after this treatment seeds won't sprout for 2 to 3 weeks. Space seeds 1 inch apart and cover ¼ inch deep with sand, vermiculite or finely sifted compost. Space rows 15 inches apart. Be sure to keep the seedbed uniformly moist.

Rooted parsley seed should be sown 1 inch apart in rows 12 to 15 inches apart.

You may prefer to start parsley using a method that is gaining in popularity. Pre-sprout seeds in a plastic bag filled with moist peat moss or milled sphagnum moss. Place the bag in a dark place at 70 to 75°F. for two weeks, then check to see if any of the seeds are sprouting. If not, check again at 17 and 21 days. At the first sign of sprouting, scatter the contents . . . moss, seeds and all . . . down a shallow furrow and cover lightly with sand or compost.

Some authors recommend planting parsley seeds direct in the garden in the fall. In zones 1 through 4, this will work if the seeding is done after the soil has frozen. However, a certain number of seeds and seedlings will perish because of weather stresses. The newer method of sprouting seeds in a plastic bag seems to be more efficient.

In the deep South and warm West, gardeners must be careful not to delay planting of parsley seeds too late in the fall. If the seeding is done just before an early and cold winter, the seedlings will grow so slowly that the plants will be too small to produce significant amounts of parsley during the winter. Then, when spring arrives with the lengthening days and shortening nights, parsley will go to seed. After that, it is of no further use. You will occasionally see a parsley plant that will over-winter and not go to seed, but it is the exception.

Planting rooted parsley too late in zones 5, 6 and 7 can result in a similar problem, that of plants going to seed the following spring before forming large enough roots for harvest.

Because of its cold tolerance, parsley makes an excellent autumn crop from direct seeding after midsummer.

TRANSPLANTING

Parsley is tolerant of cold but sensitive to heat; consequently, it doesn't make a good midsummer crop in zones 5 through 7. If intense heat usually comes early in your area, growing from transplants in the spring is an advantage because the crop can be planted early and harvested before the onset of heat. Similarly, if your seasons are quite short, such as in zone 1, starting seeds indoors and transplanting can lengthen the period of harvest considerably. To produce transplants, sow seeds indoors 8 to 10 weeks prior to the indicated direct seeding times.

Parsley seedlings need direct sunlight or strong light from fluorescent lamps if grown indoors. Without strong light, they can remain small and spindly for a long time. For this reason, many gardeners prefer to start parsley outdoors, using the pre-sprouting method because when the seedlings emerge in the cool air of springtime and are exposed to the full sun they will be heavier and stronger than those grown indoors.

Unlike leaf parsley, rooted parsley should not be transplanted because inferior root systems develop.

Thin seedlings or space transplants of leaf parsley 3 to 4 inches apart. Leaf parsley can be interplanted among other vegetables or among garden flowers. Rooted parsley should stand 5 to 6 inches apart or, if you prefer to plant it in wide bands, space the plants on 6 inch centers.

MULCHING AND CULTIVATION

Parsley responds well to mulches, either organic or plastic sheeting. Mulching is particularly recommended for the tightly curled parsley because once sand or soil is thrown up into the foliage it is difficult to wash out. Apply organic mulches when plants are half grown, to a depth of 3 to 4 inches, and pull it closely around the plants. Plastic mulches are usually employed when transplanting parsley.

Direct seeded parsley grows slowly in young stages and initial weeding should be done thoroughly, carefully and frequently. Later on, the dense growth of parsley leaves will shade out nearly weed seedlings.

WATERING

Parsley is drought resistant but will produce many more sprigs of better color if it is given a regular supply of water. If no rain falls for 10 to 14 days, apply 2 to 4 inches of water to wet the soil to a depth of 10 to 20 inches. Either furrow or sprinkler irrigation works well.

CONTAINER GROWING

Parsley is one of the best vegetables for growing in small containers. Individual plants will grow in pots and can be taken indoors in the fall for growing on windowsills. Small pots tend to dry out and blow over if grown out-of-doors; try plunging the pots in a box filled with sand, which will act as a moisture reservoir and keep the pots from blowing over.

Parsley seedlings are often planted among larger vegetables growing in containers, window boxes and planter boxes.

Rooted parsley is almost as adaptable as regular parsley to container growing, but it needs containers of 12 to 18 inches in depth to form good, long roots. Harvesting of rooted parsley is easy when all you have to do is dump out the container to get at the roots. If you wish to harvest a few roots at a time you can pull individual roots without disturbing others.

If you grow leaf or rooted parsley alone in containers, the optimum size is a 3 to 5 gallon bucket or tub. Plants can be crowded and planted on 3 inch centers. With this dense planting, you will need to harvest every other plant of rooted parsley when it is only half grown in order to allow the remaining roots to grow to full size without malformation.

ENVIRONMENTAL PROBLEMS

One of the major problems with parsley is that, because it will overwinter in all but severe climate areas, gardeners think that it should go on growing. However, the biennial plants will shoot up flowers and seed stalks. It will do no good to trim these off in the hope that the plants will resume vegatative growth. Once the hormones that govern flowering have formed in the plant, you cannot reverse the process.

Parsley seeds are notoriously slow and difficult to start. Trying to get them to come up through a crust in a heavy clay soil is almost a lost cause. Try the pre-sprouting system and plant the seeds direct in the garden. Do so early, and mark the row with a few plants of radishes so that you do not hoe out the parsley seeds and seedlings.

Small, tough, wiry roots can be a problem with rooted parsley if it is grown in the middle part of the country or in the South from spring seeding. In these areas summer comes rushing in almost before winter is over, and the roots don't have time to develop before the soil becomes dry and quite warm. For this reason, rooted parsley should be grown for fall and winter harvest in warm climate areas.

The addition of coarse organic matter to the soil should be avoided as it will cause rooted parsley to branch or fork and to become rough or hairy. It is difficult to grow good rooted parsley in heavy clay soils unless you amend them substantially with fine-grained organic matter.

In arid western soils, iron is often either deficient or unavailable to plants and results in yellowing of foliage. If this is the case in your soil, try using a slow release source such as iron chelate. It should give your parsley an intensely green color if applied in conjunction with a solution of nitrogen fertilizer.

STORAGE

Parsley can be kept in the refrigerator, wrapped in moisture proof bags for 1 week. It can be kept in a glass of water covered with a plastic bag in the refrigerator for up to 2 weeks.

Rooted parsley — To store cut off tops leaving ½ inch above the crown. Put in moisture proof bag and store at 32-40°F. May be kept several months.

FORCING FOR WINTER USE

Parsley plants can be sheared heavily, watered deeply and, after the soil drains, dug and transplanted to pots or a coldframe. Alternatively, a plastic canopy can be placed over the row, preferably of double thickness. Under a shelter such as a canopy or coldframe, parsley should be harvestable through midwinter. It will grow fairly well indoors until winter days grow short and very dark. At that point it will probably die unless it is getting strong sunlight from a south facing window. In zones 5 through 7, growing parsley under a shelter is preferable to bringing it indoors.

HARVESTING

Leaf parsley plants should be given about two months to become established and to produce several branches before you begin harvesting. Cut or snap off outer sprigs as needed to allow replacement growth to develop from the center. If, around midseason the plants lose foliage color, trim them severely, removing all the top growth, and fertilize and water heavily. This should produce a flush of new, succulent growth. Leave a few plants untrimmed to tide you over until new growth develops. Shear plants severely when you dig parsley.

PEAS

Peas

Pisum sativum, **Garden Pea, English Pea Or Green Pea**

Pisum sativum **var.** *macrocarpon*, **Edible Podded Pea, Snow Pea, Chinese Pod Pea**

Leguminosae (Pea Family)

Cool Season Annual

Killed By Heavy Frost

Full Sun

Ease Of Growth Rating, Dwarf Varieties -- 3; Tall Varieties -- 5

DESCRIPTION

Attempting to describe the taste of fresh off the vine or edible podded peas would be like trying to define love. Garden fresh peas offer a distinct advantage in taste over the canned or frozen product. This is why northern gardeners will brave blustery cold weather on the coattails of winter to plant pea seeds. Just a few days in planting time can make quite a difference in the total amount of peas harvested and the duration of pickings.

Peas are a classic cool weather vegetable. Literally, seeds must be planted in the garden as soon as the soil has dried out enough to work in the spring. Only in areas where summers are quite cool can peas be planted in late spring. While, technically, it is possible to grow a crop of peas for fall harvest in zones 3 through 7, it is not often done because of problems with mildew and seedlings perishing due to heat.

In pea descriptions you will see the terms "smooth seeded" or "wrinkle seeded". Smooth seeded varieties include all of the regular snow peas and the garden peas in the "Alaska" class. Wrinkle seeded varieties include virtually all of the standard garden peas plus the edible podded snap peas. Smooth seeds are less apt to rot when planted in cold soil. Other than that, the distinction has little significance for home gardeners.

Peas may be either in the standard garden pea class or one of the edible podded varieties. Standard garden peas are always shelled for eating; the seeds are referred to as "berries". Edible podded peas produce more vegetable matter because they are eaten pods and all.

Prior to 1979, edible podded peas were gaining gradually in popularity, riding a wave of interest in oriental cooking which employs "snow peas" in a number of ways. Then, the All-America award winning Sugar Snap edible podded snap pea was introduced. Since that time, millions of people worldwide have tried it and continue to plant it each year. Sugar Snap and its newer siblings are good not only for stir or wok frying and mixing with meats and other oriental vegetables, but also make a superb finger food and addition to salads. The tall vines of Sugar Snap are surprisingly hardy to frost and have withstood temperatures as low as 20°F. with only

moderate damage. The better dwarf snap pea varieties now available, have some resistance to mildew and certain of the pea virus diseases.

Only the dwarf pea varieties are self supporting. Their vines grow to only 12 to 14 inches in height. Two rows are often planted closely back to back so that the tendrils will entwine to keep the plants erect.

Garden peas and edible podded peas probably originated in southwest Asia. They are among the oldest cultivated vegetables and were mentioned in early Greek, Roman, Sanskrit and Egyptian writings.

SITE

When you plant garden peas very early in the spring, you can gain a week or two in maturity by planting them along the south side of the foundation of your house or where heat and sunlight are reflected into them from a solid fence. In zones 5 through 7 peas for fall harvest can be planted among the stalks of corn or okra, or in the shade of other tall vegetables. The pea vines will appreciate the cooler conditions provided by the light shade. Shade only the pea seeds planted in late summer where summers are long and hot. Spring peas are harvested early

enough to permit the planting of warm-season crops such as beans and squash in zones 2-7.

SOIL

Peas are moderately tolerant of soil acidity, and will grow best at pH levels of 5.5 to 6.8. They like sandy loam, silt loam and clay loam soils that are well drained. The plants need plenty of available calcium and magnesium for good development. Be sure to lime soils if you live in a rainy, acid soil area. Do not apply lime if you live in an arid climate where the soil pH consistently runs higher than 7 or if your garden is on a limestone based soil.

Peas remove little from the soil in the way of plant nutrients and return a substantial amount when you spade under the nitrogen-rich foliage.

SOIL PREPARATION

To develop rapidly, early spring planted peas must have good drainage and warm soil. Either build up beds or plant peas in raised beds surrounded by frames. Peas should not be given excessive amounts of nitrogen because it will encourage vine growth at the expense of pods and will delay maturity.

PLANTS PER PERSON/YIELD

Production from a given number of plants can vary by a factor of two because much depends on the duration of picking. In some areas where disease problems are minor and summers are cool, the peas may yield as many as eight pickings, but where diseases and hot weather come in early, three pickings would be more likely. The following recommendations are based on good growing conditions:

Garden Peas — Grow 50 to 60 plants per person to get 8 to 10 pounds of shelled peas.

Edible-podded Peas — Grow 50 to 60 plants for each person to get 10 to 15 pounds of pea pods.

SEED GERMINATION

Pea seeds are large and have a substantial amount of stored starch and protien to nourish seedlings. However, they are fragile and can be easily injured by crushing during shipping. Most lots of pea seeds germinate between 80 and 90 percent under favorable soil conditions.

SEED GERMINATION RATE

Soil Temp.	41°F.	50°F.	59°F.
Days to Germ.	36 days	13 days	9 days

Soil Temp.	68°F.	77°F.	86°F.
Days to Germ.	7 days	6 days	6 days

DIRECT SEEDING

Plant at the earliest recommended spring dates so that the pea crop will reach maturity while the weather is still cool. Young pea plants tolerate considerable cold, even light frosts; spring plantings are seldom harmed by cold weather.

Inexperienced gardeners often become unnecessarily concerned about losing pea seeds by planting them too early. Pea seeds will sprout in 13 days when the soil temperature reaches 50 degrees F. Gardeners can get into trouble when prolonged cold, wet weather sets in soon after planting. Under such conditions, both seeds and seedlings can rot. Some gardeners routinely treat pea seeds with fungicides as a precaution. Organic gardeners rely on good drainage and raised beds to avoid root rot or damping off disease, or they start their seeds early indoors in peat pots and transplant when the seedlings have formed four to six true leaves.

In zone 5, seeds can be planted in late summer for harvest just before the first frost. In zones 6 and 7, seeds can be planted under light shade in late summer for late winter and spring harvest or in early fall for late spring harvest. Although the seedlings of garden peas are hardy to frosts, the half-mature or mature plants can be badly hurt or killed by moderate to heavy frosts. For this reason, gardeners in frost prone areas of zones 6 and 7 usually delay planting what they call " English peas" until January, and often resort to the smooth seeded varieties such as Early Alaska which are less apt to rot in cold soil.

Direct-seed peas about 1 inch apart in rows or narrow bands. Plantings made in spring should be covered ¾ inch deep; summer and early fall sowings should be covered 1 inch deep. Earliest pea plantings are usually slow to germinate because soil temperature is still cool.

Pea yields are generally increased by treating seeds with a bacterial innoculant when planting in gardens where peas have not been sown in some time. Peas, like other plants in the legume family, have the ability to extract nitrogen from the air and fix it in their roots where it becomes a free source of fertilizer. Nitrogen-fixing bacteria are essential to this process. Some soils lack bacteria to do the job, especially new gardens. Purchase fresh powdered innoculant where you buy seeds.

TRANSPLANTING AND THINNING

Pea seeds are rarely transplanted because so many plants need to be grown for the average garden. Nevertheless, to get the jump on spring you can start pea seeds indoors three to four weeks prior to the direct seeding date. Pea plants need full sunlight or strong light from fluorescent tubes to develop compact, short plants. Harden the seedlings for about a week before transplanting them to the garden. Pea roots are sensitive to transplanting; handle the peat pots with care and be sure that the rims are covered by soil.

Thin seedlings or transplants 2 inches apart. Allow 24 inches between single rows of dwarf peas, or plant two rows 3 to 4 inches apart, with a support between the two rows, if needed. Space double row plantings 30 inches apart. Tall varieties of peas tend to billow out unless controlled by strings laced between posts.

MULCHING AND CULTIVATION

Peas are not ordinarily mulched, but a light organic mulch around late summer planted peas in warm climates will help to conserve moisture and lower the soil temperature.

Early spring planted peas do not need much cultivation, because they germinate ahead of most weeds. However, heavy soils may tend to crust and a shallow surface cultivation will help open up the soil and approve aeration. Avoid deep cultivation and avoid pulling soil up deeply around pea plants.

WATERING

Spring planted peas rarely suffer from water stress but should be watched carefully when blooms begin to form. If no rain falls from 5 to 7 days apply 1 to 2 inches of irrigation to wet the soil to a depth of 5 to 10 inches. Furrow irrigation is preferred because sprinkling tends to encourage the formation of diseases such as mildew.

CONTAINER GROWING

Peas can be grown in containers, but the production is too low per vine. Several containers would be required to produce enough for a meal.

ENVIRONMENTAL PROBLEMS

The most common problem in peas is the failure to set a large enough crop of pods to repay the time and trouble of planting. You can overcome this problem by planting adapted varieties very early and, in the extreme North, forcing the pea plantings under a clear plastic canopy or tunnel.

Another problem is the death of apparently healthy seedlings at a rather young age. This is often due to the mechanical effects of strong, gusty winds actually twisting the plants until they are severely damaged. You can prevent this by "sticking" peas with pieces of brush to which the pea tendrils can cling.

A pea malady which catches new gardeners by surprise is their habit of rapidly turning yellow and dying at the onset of hot weather. This is not due to any specific disease, but rather to a combination of diseases and the fact that peas are an annual crop and nature has programmed them to dry up and die after setting a crop of pea pods. You can avoid this by using early varieties and planting extremely early.

Heat tolerant varieties such as Freezer 692, Sugar Snap and Wando may give you a few days more production but, at this time, there is no such thing as a truly heat resistant garden pea or edible-podded pea.

HARVESTING

Peas should be harvested by grasping the stalk just below the pod and pulling the pod off without damaging the rest of the plant.

To capture the sweet, delicate taste of garden or English peas, pick just after the pods bulge, while they are bright green and velvety to the touch. Shell out the berries, which should be about one-fourth inch in diameter, by splitting the pod along the seam. Overripe pods are swollen, yellowish and spotty. At this stage, the berries will become starchy because they are beginning to dry. Pick garden peas every 5 to 7 days.

Edible podded peas (China Pea, Snow Pea) — These are eaten pod and all, and are ready when the seeds are just beginning to enlarge and the pod is still nearly flat. Pods will be from two to three inches long and brittle. Pick regular snow peas every 2 to 3 days. When overripe the pods will become fibrous and the berries of some varieties may develop an off flavor.

The edible podded All-America Selections gold medal variety, Sugar Snap, and its other snap pea relatives, are unique. They produce nearly round, thick-walled pods that snap when bent like a fresh snap bean, and can be eaten raw in salads or as finger food. The berries are plump and sweet but, since the pods can be eaten whole when quite mature, are seldom shelled. Snap peas are at their sweetest when round or nearly round in cross section. Harvest weekly.

STORAGE

Edible Pod Peas and Sugar Snap Peas may be kept in the refrigerator in a moisture proof bag up to 5 days.

Fresh peas should be kept in the pods and may be stored in the refrigerator in a moisture proof bag for 5 days.

FORCING FOR WINTER USE

Peas are not ordinarily forced in coldframes for winter use because they are somewhat day length sensitive and are shy about flowering during short days.

STORING LEFTOVER SEEDS

When placed in a sealed container with a desiccant, and stored in a refrigerator, pea seeds should germinate at least 50% after three years.

SOUTHERN PEAS

Peas, Southern Peas And Yard-Long Beans

1. **Southern Peas (Cowpea, Protepea, Field Pea, Horn Bean, Southern Table Pea, Pea)**

   Vigna unguiculata **(All Southern Peas Except The Blackeyes)**

   Vigna unguiculata unguiculata **(The Blackeyes)**

2. **Yard-Long Beans (Asparagus Beans, China Beans)**
   Vigna unguiculata sesquipedalis

Leguminosae (Pea Family)

Warm Season, Heat Resistant

Annual

Killed By Light Frosts

Full Sun

Ease Of Growth Rating, Bush Peas -- 3; Climbing Yard-Long Beans -- 4

DESCRIPTION

Southern peas and yard-long beans are among the best vegetables for warm, humid climates and, to a lesser extent, warm, dry climates. They are a staple food item for many southern families and are grown in large amounts. Southern peas are better known than their "kissing cousin", yard-long beans, which are often treated more as a curiosity than as the productive, dependable vegetable they really are.

Southern peas suffer identity problems. Although botanically much closer to garden beans, they have, in the Deep South, been called "peas" ever since their introduction in colonial days. They are listed under "Peas" in southern catalogs and in bulletins from State Cooperative Extension Services. While granting their similarity to beans, this book will bow to popular usage and list this vegetable under "Peas".

Southern pea and yard-long bean plants superficially resemble those of snap beans but have stiffer stems, glossier foliage and a habit of holding pods horizontally and splayed out, like the fingers on a hand. They are grown for their long pods which are shelled to produce green peas. Any immature pods which are picked by error can be snapped like bean pods and thrown into the cookpot with the green seeds. Overly mature pods can be picked and spread out to ripen and produce dry seeds for storage.

Three distinct growth habits exist: the upright bush or bunch plants, the semi-vining plants which adapt to growing among corn rows for support, and the high climbing vines of the yard-long bean.

Southern peas and yard-long beans are thought to be native to southern Asia (probably India). From there they spread across Africa, where they natural-ized, to southern Asia and adjacent islands, and to the Mediterranean region of Europe. The culinary varieties were brought to Jamaica from Africa in 1675 and reached Florida via the West Indies in 1700. In 1714, they were mentioned as being grown in North Carolina.

In the American Colonies, the southern pea was first known as "callivance" and later as "Indian pea", "southern pea", "southern field pea" and "cornfield pea". The name "cowpea" referred to a rank-growing variety that was grazed or cut and used for hay.

When a southerner says "peas", he or she is almost always referring to southern peas. "English peas" in the South are green peas, *Pisum sativum*.

SOIL

Sandy, sandy loam or fast-draining clay loam soils within a pH range of 5.5 to 7.0 are best. However, since so many soils in the south are clays, they are adapted to growing southern peas by ridging up rows or building raised beds. Southern peas need good levels of available calcium and magnesium but liming should be done only when indicated by soil analysis. Excessive liming can induce deficiencies of certain nutrient elements; this is most apt to occur when application rates exceed 1 ton per acre (5 lbs. per 100 sq. ft.).

SITE

Grow southern peas and yard-long beans in full sun and well away from previous crop sites, to reduce disease problems. Good drainage is essential to healthy plant growth.

SOIL PREPARATION

Prepare soil by working organic and mineral additives soil tests. Calcium and phosphorus move down in the soil very slowly; surface applications are not efficient. Except on sandy, well drained soil, build up ridged rows or raised beds. Although southern peas and yard-long beans will grow fairly well on rather poor soil, they will produce heavier crops if a light to moderate amount of organic matter is worked into the rootzone, along with generous amounts of phosphate and potash. If you have never before gown these vegetables in your garden, buy a specific bacterial "inoculant" for southern peas (it will also do for yard-long beans). See "Beans" for an explanation of how the nitrogen fixing bacteria in the inoculant help extract nitrogen from the air.

PLANTS PER PERSON/YIELD

In zones 5-7 where southern peas are a staple food, 50 ft. of row per person is standard, where garden space permits. This should give you 1 to 2

cups of shelled peas per picking. With multiple pickings you should harvest about 40 lbs. of shelled peas per 100 ft. of row over the life of the crop.

Elsewhere and in small southern gardens, the plant habit enters into the decision of how much to plant:

Bush varieties: Grow 15 to 18 plants per person to get 12 to 15 lbs. mixed shelled peas and snappers.

Semi-vining and vining varieties: Grow 10 to 12 plants for each person to produce a total of 15 to 20 lbs. of mixed shelled peas and snapped pods.

Pods are generally picked twice a week. If you harvest too few for a mess of peas, keep them in the vegetable crisper and combine with the following picking rather than let the pods hang on the vines and create a drain on nutrients.

SEED GERMINATION

Germination averages 50-60 percent and does not hold well. Technically, southern pea seeds will germinate in the range of 50-90°F., but at lower soil temperatures germination is slow and many seedlings are lost to damping off. At the high range of soil temperatures, the heat at the surface of the soil and just above it can be injurious to seedlings. A more reasonable soil temperature range is 60-75°F., with a 65° minimum recommended by most States. Seeds sprout in 7-10 days.

DIRECT SEEDING

Southern peas require warm soil to germinate satisfactorily, and even then stands are often sparse. Sow seeds thickly, ½" deep when planting in late spring and 1" deep when planting in late summer for fall harvest. On bush varieties, make rows 2 feet apart or plant in double rows 18 inches apart. Give each plant in double rows maximum space by making sure that the plants in one row are not directly across from those in the second row.

Semi-vining or vining types — plant 2 or 3 seeds to produce one seedling by each support. When trellising, plant seeds 3" apart.

TRANSPLANTING AND THINNING

There is no advantage to transplanting southern peas or yard-long beans.

Thin seedlings of bush varieties to stand 6 inches

apart and semi-vining or vining types 12 inches apart.

MULCHING AND CULTIVATION

Do not mulch southern peas unless you are growing them on sandy soil which tends to become extremely hot. Under these conditions you can apply a shallow layer of organic mulch under the plants to keep the soil cool and to conserve moisture. If you mulch too deeply, you may cause the plants to become growthy and to produce fewer pods.

Weeds grow well around southern peas and asparagus beans because of the nitrogen they fix from the air. The plants are fairly deep-rooted, and you do not have to be super-careful when hoeing around them. Hoeing should be done by scraping rather than chopping.

SUPPORTING STRUCTURES

Bush varieties need no support. The vines are erect and stand up against most storms.

Varieties with short runners grow well when interplanted among corn, or you can make a simple trellis by nailing a 2 ft. wide strip of chicken wire to posts 2-3 ft. high. If even a few runners twine around the chicken wire, they will keep the plants from falling over. You can let the short-runner varieties sprawl on the ground but this complicates picking.

Yard-long beans have such long, twining runners that they need teepees or trellises 6 to 8 ft. high, as high as you can reach to pick pods. The vines can grow heavy by the end of the season; if trellised they will need strong strings.

WATERING

These plants are relatively drought resistant, but substantial increases in yields can be realized if water is supplied every 10 to 14 days between rains. Apply 1"-2" of water to wet the soil to a depth of 5" to 10". Furrow irrigation is preferred because it does not wet the foliage and it lets water soak into clay soils.

CONTAINER GROWING

Southern peas are not well suited to container culture because of the number of plants required to produce a meal. However, yard-long beans make a decorative and practical container plant. Grow 4 or 5 plants around the perimeter of a 20-30 gallon container and train runners up long strings to a center post. One or two such "pea trees" should yield enough pods to feed a small family at a picking. Container soil will need to be inoculated with nitrogen-fixing bacteria for healthy growth of southern peas.

ENVIRONMENTAL PROBLEMS

Some problems with southern peas and yard-long beans come from planting them in marginally adapted areas. Home gardeners are not entirely at fault; they take at face value the "days to maturity" listed in seed catalogs, not realizing these are based on readings taken in zones 5-7. At the northern limits of adaptability, add two weeks to the stated maturity dates, because of the slow accumulation of the heat units necessary to maturity.

DISEASES

At one time it became extremely difficult to grow southern peas in zones 5-7, particularly east of central Texas and Oklahoma, because of the prevalence of root-knot nematodes, fusarium wilt and pea virus diseases. Yields of fresh shelled peas per acre dropped from 5,000 lbs. to as little as 350 lbs., and zero in some extreme cases. Southern plant breeders stepped in and developed several excellent varieties with multiple disease resistance, good flavor and high production. While they were at it, they made shelling easier. When reading descriptions of southern pea varieties, such disease resistance is one of the first things to look for, if you live in zones 5-7.

HARVESTING

Experienced pea growers go by the change in pod color, sheen of the pod skin and the relative lumpiness when deciding which pods to pull. Typically, a mess of peas will contain pods that are overly mature as well as some which are too green to be shelled. Go ahead and shell the overly mature pods if the seeds are still soft. If the seeds are hard, lay the pods aside to dry for shelling later. The young, immature pods, called "snappers", can be snapped like beans and tossed into the pot. Try to keep all the mature pods picked off; if you let a heavy set accumulate, it will reduce the vigor of the vines and they will be slow to repeat. To avoid breaking off branches hold on to the fruiting stem with one hand while pulling pods with the other.

STORAGE

Years ago, surplus southern peas were canned or dried. Now, most of them are frozen.

While accumulating enough peas to freeze, you can hold them in the shell for about 5 days. When you strip the vines before the first fall frost you will harvest quite a few snappers. These can be mixed with shelled seeds after freezing. Snapped yard-long beans can be handled similarly.

Dried seeds of unusual varieties are impossible to buy in food stores. To save your own, pick pods when they are half dry. If you leave them on the vine they may split and shatter or spoil from rain. Spread the pods out in a warm dry place until they are dry enough for the pods to crumble in your hand. Shell and winnow out the chaff. The old way of killing weevil seeds in seeds before storage was to heat them in an oven. Now, dried seeds are placed in a freezer, both to preserve them and to eliminate weevils and other storage insects affecting stored seeds.

STORING LEFTOVER SEEDS

In past years, home-saved seeds of southern peas dropped drastically in germination when stored for more than one season. Now, gardeners are preserving germination by keeping fully dried seeds in the freezer. After normal air-drying, seeds for planting can be further dried by sealing them in a jar with a desiccant such as powdered milk or silica gel. Keep the container in the shade at room temperature for 2 weeks for the desiccant to do its work, remove it, reseal the container and store in the freezer. Seeds thus prepared should germinate 50-60 percent after 2 years of storage.

PEPPERS

Pepper, Sweet And Hot

Capsicum annuum, Including All Garden Varieties Of Sweet, Hot And Ornamental Peppers

Capsicum annuum Var. *minimum*, The Bird Pepper (Chili Piquin), The Tree Pepper (Chili Abore), And Other Small-Fruited Types

Capsicum frutescens, The Tabasco Pepper

Solonaceae (Nightshade Family)

Warm Season Perennials, Killed By Light Frosts

Full Sun Or Partial Shade

Ease Of Growth Rating, Sweet Peppers -- 4; Hot Peppers -- 3; From Plants You Grow From Seeds -- 6

DESCRIPTION

Peppers are one of the most valuable crops you can grow in your home garden or in containers because the fruits are usually expensive when purchased and, under garden conditions, the plants can produce continuously for weeks or months. Surplus or end-of-season peppers can be diced or sliced for freezing, minimizing waste.

Peppers are produced on bushy plants which may range in height from 14 to 48 inches, depending on the variety and the length of the growing season. The fruits on some varieties are upright while others are pendent.

Only a few years ago it was difficult to find pepper varieties which would mature reliably in the northern states during a cold, moist summer. Back then, peppers would set fruit within a narrow range of temperatures — approximately 65 to 80°F. Now, however, vigorous hybrids have been developed that set fruits at 60 to 85°F., given adequate moisture and balanced nutrition. Also, the effects of a widespread and serious scourge of peppers have been reduced by the introduction of varieties resistant to tobacco mosaic virus, a crippling disease of peppers.

Bell and hot peppers of the many different sizes and shapes may be either red or yellow when ready for harvest, depending on the variety. All peppers will turn red when ripe, even the yellow or golden varieties. At the red stage, sweet peppers are especially delicious, plump and sugary.

Many of the dried and pickled pepper preparations for sale on spice and condiment racks come from **C. annuum** varieties: pimentos (also called pimientos), paprika, chili and cayenne powder and pepper sauce, for example. Only tabasco sauce comes from the distinct perennial species, the pungent **C. frutescens.**

Bird and tree peppers are cultivated by Mexican-American families in the Southwest and plants have escaped to grow wild in fields and under light forest cover in parts of zone 7. They bear small, round or thimble-shaped fruits which are fiercely hot. Dried whole bird peppers are sold in Southwestern stores along with dried jalapeno and chili peppers. Seeds of bird and tree peppers are rarely sold but are usually passed around among friends and families.

Other species are grown in South and Central America and the Caribbean but are rarely seen in the USA, because of their need for a long, warm growing season, similar to that of tabasco.

Botanically, peppers present a confusing picture, with shirt-tail relatives crashing family reunions. Ornamental peppers, long considered a pot plant, are now being planted in vegetable and flower gardens in zones 2 through 7. These heat-resistant peppers grow on small, neat plants. All are edible and, depending on the selection, range from barely warm to the tongue, to very hot. Although ornamental peppers are edible, they are listed among the flowers in seed catalogs and seed racks.

Ornamental peppers are replacing the poisonous "Jerusalem cherry", sometimes called "Christmas cherry", a distantly-related solanum with round yellow or orange-red fruits.

Garden peppers are not related to black pepper, *Piper nigrum*. Capsicum, as the garden pepper is known in parts of Europe, originated in tropical America and was taken back to Europe by early explorers. The five botanical groups with the species **C. annuum** can be accidentally or hand-crossed and selections from such crosses are rather simple to increase. Consequently, thousands of unnamed pepper varieties may be in existence, mostly in hot peppers grown from seeds saved by people in certain villages or family groups.

SITE

Grow peppers in full sun. In all zones, but especially 1 and 2, sandy soil helps to bring early maturity. Reflected light and heat from a wall or fence also helps. In humid areas of zones 6 and 7, peppers will grow about a foot higher than listed in descriptions because of the long, favorable growing season.

If you have a large garden, plant peppers away from cucumbers, eggplants and tomatoes. Tobacco mosaic virus can be transmitted among members of the solanum family but cucumber mosaic virus is more democratic; it attacks plants in other species as well as vine crops.

SOIL

Peppers are one of the few crops for which it would pay you to build an "ideal" soil, if yours does not already meet the specifications. Peppers prefer a sandy or sandy loam soil, well fortified with organic matter to act as a reservoir for water and plant nutrients. A well-drained, warm soil can produce earlier, heavier crops. Peppers will grow well at a pH range of 5.5 to 7.0 and use only moderate amounts of plant nutrients.

SOIL PREPARATION

Peppers are a long season crop and, on other than sandy, well-drained soil, benefit from raised beds or ridged-up rows. The plants have an extensive root system which is easily harmed by trampling around the plants. Raising beds or ridging up rows discourages foot traffic around the plants. Make wide beds or broad ridges; narrow, high beds dry out too fast and are difficult to water without erosion. Work in 1 to 2 inches of well decomposed compost, packaged pasteurized manure or peatmoss; use soil excavated from the foot paths to gain elevation on the beds.

Peppers need substantial amount of magnesium to grow well. This nutrient is usually in good supply where soils are limed with dolomitic limestone. But, limestone is not needed in the arid West and on limestone-based soils elsewhere. In these areas, magnesium can be supplied through applications of epsom salts, which is magnesium sulphate. Incorporate 5 pounds per 100 square feet or, preferably, at rates recommended by soil testing.

PLANTS PER PERSON/YIELD

Grow 3 to 4 plants of sweet peppers per person to get 36 to 40 fruits of the large bell types and 50 to 60 of the smaller salad or pickling types.

Three to four plants per person of hot peppers and small-fruited sweet peppers can yield up to 150 fruits.

SEED GERMINATION

The Federal Minimum Standard for pepper seed germination is 55 percent. However, most home garden seed lots will germinate in the 65 to 70 percent range.

Being from a tropical plant, pepper seeds need warm soil to germinate quickly; a soil temperature of 70 to 85°F. is ideal:

Temp. Range	50°F.	59°F.	68°F.	77°F.
Days to Germ.	No Germ.	25 days	12 days	8 days

Temp. Range	86°F.	95°F.	104°F.
Days to Germ.	8 days	9 days	No Germ.

Pepper plants should be grown at temperatures of 75°F. days and 65°F. nights.

DIRECT SEEDING

Peppers are almost always transplanted for home garden use and not seeded where they are to remain. The seeds need quite warm soil to germinate, and seedlings grow slowly. Tiny seedlings can be swamped by weeds or killed by organisms in garden soil. An exception is the practice of outdoor seeding in midsummer in zones 6 and 7. There, outdoor nursery beds can be shaded to keep the soil from drying and the shade gradually removed when the seedlings have emerged. Treating seeds with a fungicide for outdoor seeding will generally improve germination and survival. The seedlings are transplanted from the nursery bed to the garden.

TRANSPLANTING AND THINNING

Seven to eight weeks prior to the indicated outdoor transplant dates (8 to 10 weeks in zones 1 and 2), sow pepper seeds indoors. Cover ½ inch deep with vermiculite or milled sphagnum moss. Keep the seed flat temperature in the mid-70's with a heating cable or a grounded incandescent light bulb burning beneath the table, shielded from drips.

Pepper plants should not be set in the garden until the soil has warmed. A sheet of clear plastic laid over the pepper bed 2 to 3 weeks prior to transplanting will trap heat from the sun and raise the soil temperature five to ten degrees over that of surrounding soil. In zones 1 & 2 transplant through the clear plastic; see "Mulching" for other zones.

Protect pepper seedlings from wind and cutworms with bottomless plastic jugs placed over them at planting time. These jugs will give protection from light frosts.

Spacing of pepper plants depends partly on the variety and partly on the planting zone. Compact varieties can be spaced 18 inches apart, except in

zones 5 through 7 where 24 inch spacing will minimize crowding. Hot peppers and vigorous, late sweet pepper varieties can be spaced 24 to 30 inches apart, in rows 24 - 30 inches apart. Space tabasco peppers 36 x 36 inches. Gardeners usually interplant fast-growing salad crops among young pepper plants to keep down weeds and to utilize the open soil between plants.

MULCHING AND CULTIVATING

Peppers respond positively to mulches. In the far North, clear plastic mulch is recommended to raise the soil temperature to the maximum extent while further south, black plastic is preferred. By warming the soil early in the season you speed up the release of nutrients and the formation of feeder roots.

Organic mulches are preferred in zones 6 and 7; they are applied in midsummer to cool the soil during the hottest part of the year. Organic mulches can be used in zones 3 through 5 as well, but they should not be applied until late summer. Then, they serve to retain heat absorbed during the summer and to keep the soil temperature somewhat higher during fall when night temperatures are dropping steadily.

Avoid cultivating peppers deeply. Careless hoeing or using a tiller to create a dust mulch near pepper plants can cause severe root pruning and drying out of the surface soil, both of which can reduce production and delay harvest.

SUPPORTING STRUCTURES

Even the compact varieties of peppers are prone to falling over during rain and windstorms when loaded with fruit. A single stake per plant will generally suffice. Three or four foot stakes are best; the spare length can be used to tie up the plant as it grows. Hot pepper plants are better about not falling over but, near the end of the season, may need staking.

FERTILIZING

Too much nitrogen can cause vegetation growth at the expense of fruits.

Peppers can benefit from side-dressing or liquid feeding when the plants are rather small and again when two or three fruits have set. If plants are starved when young, production will drop significantly.

WATERING

Of all the vegetables in your garden, peppers will respond best to drip irrigation in greater production over a longer period and with a higher percentage of uniform fruit. With conventional irrigation, peppers should be given two to four inches of water every 7 to 10 days if no rain falls. This will moisten the soil to a depth of 10 to 20 inches. Pepper foliage is a good indicator; a certain amount of wilting is normal for hot, dry days but the plant should plump up by

nightfall. If not, they are being stressed for water and the production will suffer. Water more frequently. Occasionally you may see peppers suffering from blossom-end rot, caused by uneven water levels in the soil, combined with a deficiency of available calcium.

CONTAINER GROWING

Peppers are among the best vegetables for growing in containers, in fact, could be rated with tomatoes for a combination of productivity, sustained yield and money value of the crop. Plant one pepper seedling per three gallon container or 2 to 3 per 5 gallon container. Shallow containers work best for peppers, for example, foot-tubs with a 3:1 diameter-to-depth ratio. Peppers will grow in boxes if the soil is at least a foot deep.

Fortify standard potting soils with 1/4 cup of dolomitic limestone per gallon of mix to insure adequate amounts of calcium and magnesium throughout the growing season. Low nitrogen soluble fertilizers may be hard to find; one in the 5-10-10 bracket may be found among the fertilizers labeled for flowers. Use it; the peppers won't know the difference.

ENVIRONMENTAL PROBLEMS

Some pepper problems are due to the use of unadapted or disease-susceptible varieties. Home gardeners often choose varieties with large fruit, not realizing that these usually mature several days later than smaller fruited types and may be more finicky about temperature and soil moisture.

A complaint occasionally heard is "sweet peppers turning out to be hot." First off, sweet peppers and hot peppers rarely cross because the blossoms are self pollinating. Even then, you would have to save the seeds and plant them next year to see or taste the results of the cross.

Mixups do occur in growing and merchandising plants, and customers can be careless about pulling out and replacing labels. In general, sweet pepper varieties have larger, broader leaves than hot peppers.

When buying plants of any pepper variety, be careful to select packs or flats in which the label does not seem to have been moved or tampered with. If the label does not list a variety name, don't buy it. A grower who will label plants only as "Peppers, Bell", for example, has a low opinion of your intelligence and little concern for your gardening welfare.

A number of mosaic virus diseases of peppers are common across much of North America. The most prevalent, tobacco mosaic virus, can be transmitted either by the feeding of aphids or mechanically by picking from an infected plant and then a clean one. This virus can be carried on the hands of smokers. Curing tobacco for cigarettes does not kill the virus.

The importance of using plants resistant to this virus cannot be over-emphasized. Look for the acronym TMV in seed catalogs and on seed packets.

HARVEST

It is true that red ripe peppers are delicious . . . sweet, crunchy and vitamin laden. Yet, you should not leave many fruits on the plants beyond maturity, the reaching of full size. During the two to three weeks required for fully grown fruit to ripen, they will be filling up with seeds and drawing on the plant for nutrients. This stresses the plant and reduces production.

Pepper stems are brittle and easily damaged; harvest peppers by carefully cutting or snapping the fruit from the plant, along with a short piece of stalk.

Sweet peppers can be harvested any time they reach a size you want. Most bell varieties stop enlarging after reaching a length of about 4 inches.

Hot peppers are best after they reach full size and develop their normal mature color which may be green, red or yellow, depending on the variety. Hot peppers produce a volatile oil that irritates the mucous membranes. Avoid its unpleasantness by harvesting in the cool part of the day and by wearing rubber gloves during harvest and handling. Wear glasses or sunglasses during harvest to remind yourself not to rub your eyes with hands coated with hot pepper juice.

When a light frost is forecast, you can cover pepper plants to avoid damage; this might protect the pepper plants through Indian summer and give you a bonus of 2 to 3 weeks of production. When temperatures below 26-28°F. are forecast, dig the vines, roots and all, and hang them under shelter. Pick off and freeze the fruits as soon as possible.

STORAGE

Peppers can be stored in the refrigerator in a moisture proof bag for 1 week. To prepare sweet peppers for canning, freezing, drying or pickling: Wash, remove stem and seeds.

Hot pepper plants can be uprooted and allowed to dry or the pods can be taken from the bush and strung on a string to dry. When dry, they should be stored in a cool, dry, well-ventilated place.

STORING LEFTOVER SEEDS

Seeds placed in a sealed container with a desiccant and stored in a refrigerator should germinate about 50 percent after two years. Pepper seeds have only average longevity.

POTATOES

Potato

Irish Potato, White Potato

Solanum tuberosum

Solanaceae (Nightshade Family)

Cool Season

Grown As An Annual

Killed By Heavy Frosts

Full Sun Or Partial Shade

Ease Of Growth Rating, From Tubers -- 4; From True Seed -- 6

DESCRIPTION

Potatoes grow as underground tubers on medium-sized, bushy plants with succulent stems usually called "vines." The tubers form on the stem between the seed or seed piece and the surface of the soil. Potatoes are widely adapted but are most productive in areas that can provide 4 months of relatively cool weather free of light frosts. Potatoes can be grown from pieces of tubers, confusingly called "seed pieces," or from transplants grown from true seeds. Seed stores and farm oriented garden centers stock bags of seed potatoes in the spring, but they are seldom available until after the coldest weather is past.

Potatoes can be started from true seed, but this way has its limitations. Only a few varieties of potato are available from seed at this writing, and they are somewhat inferior in production, size, and quality to potatoes grown from pieces of tubers. Breeders are working to develop comparable seed grown varieties, because true potato seeds are easy to store and ship at any season of the year. Seed is a reliable way of starting plants for mid to late summer planting in zones 3, 4, 5, 6, and 7 when tuber pieces are seldom available. Scientists are interested in true potato seed since, ordinarily, plant diseases are not transmitted through seeds to the next generation. Also, potato seed offers potential in short season areas where seedlings started indoors can be transplanted in the garden once the danger of heavy frost is past.

Wild potatoes can still be found in Peru, where they originated. Early records of potatoes in Mexico indicate that country is another possible point of origin or perhaps a way station for potatoes brought from South America. Early Portugese explorers of the Andes reported finding egg-sized potatoes in yellow, white, and purple colors. Other colors also occur in nature, including black, blue, pink, and piebald.

The transformation of potatoes in the 450 years since their discovery by Europeans is astounding.

The wild species develop extremely large, late fruiting plants when taken to low altitude northern temperate climates. From these unadapted plants have come our modern compact-vined, early, extremely productive varieties. Our plant breeders did not propagate the wild potatoes of exotic colors, although the yellow fleshed potato is fairly popular in parts of Europe and a blue potato is sold as a novelty.

SITE

Place your potatoes in full sun and near a water source. In all probability, furrow or drip irrigation will be needed. Light shade can be tolerated but will decrease the yield except in zones 6 & 7. Do not plant in soil used for growing potatoes during the past 3 years, since some diseases from the refuse of previous crops can survive in the soil.

SOIL

Potatoes grow best in sandy loam or highly organic soils. In heavier soils it is better to grow them *on top of* the soil rather than *in* it. Potatoes are one of the few crops that are very tolerant of soil acidity. They will grow at pH levels of 5.0-6.8, but their susceptibility to "scab" disease increases at pH levels of 6.5 and above. If your soil is only slightly acid, don't lime it. Instead, add agricultural gypsum as a source of calcium and sulfur. Very acid soils, limed every 1-3 years, are unlikely to foster potato scab disease. Organic soils such as muck are ideal for potatoes if the pH level is kept below neutral (7.0). Few other crops require as much of the major nutrients as potatoes, which is understandable in view of the weight of tubers produced in a relatively short growing season.

SOIL PREPARATION

Experienced potato growers generally avoid adding large amounts of fresh stable or cattle manure to potato beds because it tends to raise the soil pH and encourage potato scab disease.

PLANTS PER PERSON/YIELD

Grow 15-20 plants for each person. Depending on the variety and growing conditions, this should yield 25-40 pounds of full sized potatoes, enough for fresh use and storage.

SEED GERMINATION

True potato seed has only recently entered the home garden market in substantial amounts, and germination data has not been quantified. However, it is known to sprout well at 70-75°F. and seeds should germinate around 60%. However, shortly after sprouts appear the growing temperature should be dropped to around 65°F. to produce stocky, compact seedlings.

PLANTING SEED PIECES

When planting tuber seed pieces, every seed piece should grow if it has been properly cut to have at least 2 eyes. As commercial growers have long known, large seed pieces produce somewhat stronger plants in the early stages. Precut seed pieces or smaller pieces called "eyes" are sometimes delivered on mailorders; they will have been disinfected and treated with a fungicide. If you cut your own seed pieces, remove the potatoes from cool storage about a week prior to cutting. A single tuber will usually produce 4-6 pieces, each with 1 or 2 eyes. Plant the pieces as soon as possible after cutting, certainly within 7 days. Dusting with a fungicidal seed treatment is recommended, but organic gardeners usually prefer to waive this treatment. A simple way to dust seed pieces is to measure a teaspoon of fungicide per dozen pieces and shake them together in a paper bag. The seed pieces should be dry for the dust to cling evenly.

Plant the seed pieces 2-3 inches deep and 9-12 inches apart. Rows should be 32-36 inches apart. Subsequent pulling up of the soil around the plants should cover the tubers to a total depth of 5-6 inches. In zones 6 and 7 tuber pieces have a distinct advantage over seeds for early spring planting. Pieces can be planted 4-5 weeks before the last expected frost and covered with 2-4 inches of porous soil. The roots will proliferate weeks before the sprouts emerge and when the soil warms the plants will grow quickly, drawing on the established root systems.

If you purchase precut seed pieces, avoid those where the buds have sprouted beyond ¼-½ inch length, because these can be damaged in handling and planting. Specify "certified" seed pieces if available; they usually come in packages containing 50 pieces but sometimes can be purchased in bulk by weight. These tubers come from plants known to be free of potato mosaic and grown in seed fields inspected for other serious potato diseases.

WATERING

Potatoes grow best in a climate with frequent but not excessive rains. Never permit plants to wilt — especially once they have begun to blossom. Even a temporary shortage of soil moisture during the time of tuber formation can cause cracked or malformed tubers. Irrigate potatoes if no rain falls for 5-7 days. Apply 3-5 inches to moisten the soil to depths of 15-25 inches. If you have to irrigate frequently, watch the leaves of your crop for yellowing, a sign of nitrogen deficiency, because heavy watering can leach away this important nutrient.

CONTAINER GROWING

Potatoes are easy to grow in containers. Plant 1 seed piece or seedling per 3-gallon container or 2 per 5-gallon bucket or plastic bag. You can plant potatoes very early in containers and grow them under protection to anticipate the season. Or, at the end of the growing season, they can be brought under protection to gain a few weeks more growth, and produce more and larger tubers. Keep in mind the potato plant's high requirement for water. You may have to set the container in a partially shaded area away from drying winds to prevent wilting of foliage in midafternoon. Container growing of potatoes is not a practical way to produce a significant number of tubers, of course. However, potatoes do make a novel and decorative plant for container culture.

TRANSPLANTING AND THINNING

Several seed companies sell the true potato seeds. These should be sown indoors 5-6 weeks prior to the indicated transplant time. Be careful not to start seeds more than 8 weeks before the frost free date because seedlings can form tubers in the pot or flat after 10-12 weeks, and this premature fruiting will slow vegetative growth. To start the seeds, cover them only 1/16-1/8 inch deep in a seedling flat or pot. When the seedlings have developed 4-6 leaves, move them to individual peat pots. To develop strong, stocky plants, sprout and grow the seedlings in full sun or under strong light from fluorescent lamps. Harden the transplants and do not disrupt the root system when planting in the garden. Set the transplants in the garden at the spacing recommended for seed pieces. (See PLANTING SEED PIECES).

Thin plants to 9-12 inches apart in rows spaced 30-36 inches apart. As many as 2-3 sprouts may emerge from a single seed piece, but these should be considered as a single plant and not thinned.

MULCHING AND CULTIVATION

Potatoes are well suited for mulching. The principle behind mulching potatoes is to keep the soil moist but not hot, for tubers will not set at soil temperatures over 84°F. Spring crops in zones 1, 2, 3, and 4 can be mulched with black plastic. Spring crops in zones 5, 6, and 7 are not usually mulched with plastic since it tends to accumulate too much heat towards the end of the growing season. In zones 1, 2, 3, and 4, organic mulches can also be used, but not until about the time potatoes begin to bloom, so that the soil is warm enough for fast plant growth.

Up until the time of blooming, cultivate by pulling soil up around the plants and, in the process, killing the weed seedlings. When you see the first blooms, cultivation should be discontinued, because at that time tubers are beginning to form and it is important to avoid root disturbance.

Gardeners in zones 6 and 7 who like to plant a second crop of potatoes in the fall often have

POTATOES

difficulty in locating potato seed sources, because the new crop of tubers will not have been harvested and prepared for shipping at this early date. Consequently, they plan ahead and save some of their own spring crop of tubers for cutting into seed pieces for their fall crop, or they start transplants from seed.

Days to maturity for potato varieties depends largely on soil conditions, levels of soil fertility and moisture, and freedom from major diseases. Commercial potato growers sometimes will artificially terminate the vegetative growth of their crops to harvest in time to meet harvest schedules. For purposes of garden planting, you can expect early varieties to mature in about 90 days, medium early to midseason in about 100 days, and late varieties in 110-120 days from planting.

ENVIRONMENTAL PROBLEMS

Failure to set tubers, or a shy set, may be due to four causes: planting too late, feeding too stingily, stressing plants with dry soil, or planting an unadapted variety. Rotting of seed pieces may be caused by failure to disinfect with a dust or dip of fungicide, by failure to plant soon after cutting, or by the use of dried-out seed pieces that were cured and held at a low humidity level.

Growers in zones 6 and 7 often have problems with poor set of tubers or poor growth of plants planted in mid-spring. This is due to high soil temperatures when plants are beginning to set tubers, and can be overcome, to some extent, by early spring planting or by delaying planting until the late fall. In zones 6 and 7 high soil temperatures have been counteracted in late summer crops planted in sandy soils by using deep mulches of straw, hay, or pine needles to reduce the soil temperature. Mulching is usually combined with overhead watering, but when plantings are made in late summer, foliage diseases can be troublesome. Foliage diseases can be reduced and periodic waterings simplified by the installation of a drip irrigation system or by inverting a soaker hose down the potato bed for weekly waterings.

HARVESTING

Early harvests of tender skinned new potatoes can be made shortly after the plants flower, but the main harvest of potatoes for storage should wait until the tops yellow and wither. At this stage the underground tubers are fully developed. To harvest, dig potatoes carefully, working in toward the center from a foot or so beyond the plant. A flat-tired spading fork works best. Shake the soil as you pry up and you will work most of the clustered tubers to the surface of the soil. Allow tubers to dry for an hour or so on top of the ground, and store only those which are not bruised or cut. Use any damaged tubers as soon as possible. If the tubers dug from your first trial plant are small and the tops are still green, delay harvesting until the tops yellow and die, but harvest before a frost.

STORAGE

Fall harvested potatoes can be stored in a cool, dark place for 6 months. New spring and early summer potatoes are more perishable; store in a cool dark place but use within a week. Cure fall harvest of potatoes for storage immediately after harvest. Try holding potatoes at 50-75°F. with relatively high humidity for 10-14 days. After curing, store potatoes at 35-40°F. in a dark, ventilated area; they require moderate humidity but moisture should not collect on them. Only potatoes without serious cuts or bruises should be stored. Exposure to light during storage will turn the skins green. Potatoes stored for long periods of time at refrigerator temperature take on a sweet flavor since some of their starches will turn to sugar.

Sprouting can be expected if potatoes are stored at 40°F. or above for 2 or 3 months. Store new potatoes by dusting them with quicklime (hydrated). Place them on boards or screening, one layer deep and not touching one another. They will remain dormant for about 3 months after they mature, without sprouting. If certified tubers are unavailable for the fall crop, hold over some sprouting tubers from the spring crop to use as seed pieces.

STORING LEFTOVER SEEDS

True potato seeds, when stored as recommended, should germinate at least 50% after 2 years.

RADISH

DESCRIPTION

No other vegetable except cress is ready for eating as soon after planting as salad radishes. Spring radishes are usually ready a few days before spinach. At one time, radishes were among the most popular vegetables for direct seeding in the garden, but have dropped slightly in the ratings.

Basically, two types of radishes are available: those grown principally for salads and the larger

Radish

Raphanus sativus

Cruciferae (Mustard Family)

Cool Season Annual, Killed By Heavy Frosts

Full Sun Or Partial Shade

Ease Of Growth Rating - Salad Radishes -- 2; Winter Radishes -- 3 types that are often cooked. The latter are referred to as "winter radishes" or in Japan as "Daikon". Winter radishes are much larger and later maturing than salad radishes.

Radish tops are of various heights, depending on the variety, and roots can range from the size of a marble to more than 30 pounds in weight. This variation in size of roots and tops dictates differences in garden site and culture.

Radishes are native to the temperate regions of the Old World. Perhaps due to their ease of culture, radishes have long been grown. Egyptians, Chinese and Japanese have cultivated radishes for centuries and many types are grown, some occasionally weighing 15 to 20 pounds each. This weight is far surpassed by a radish described by Matthiolus in 1544, as weighing over 100 pounds.

SITE

Plant in full sun. In zones 5 through 7 radishes planted for fall harvest benefit from afternoon shade. Spring radishes grow so rapidly that they are usually interplanted among other vegetables such as spinach, lettuce or garden peas. Planting them alone in a special bed wastes garden space. Summer or fall planted radishes of the large varieties for storage need to be in special beds because of their size.

SOIL

Radishes are moderately tolerant of acid soil and grow best at pH levels of 5.5 to 6.8. Salad radishes mature in only 3 to 4 weeks from seed planting, and will grow on soils that are too poor to sustain longer season crops. Spring sown radishes are ready faster when grown on sandy or silt loam soils. Late radish crops like the higher moisture and fertility levels that can be maintained on heavier soils.

SOIL PREPARATION

Radishes should be grown without "checking" of their development. To get them off to a fast start and to supply the nutrients and water needed to keep roots growing steadily, work in 1 to 2 inches of well rotted manure or compost. Radishes are often interplanted among cole crops such as cabbage, broccoli and cauliflower. The same soil preparation is appropriate for cole crops and radishes. Radishes should be planted rather thinly; once growth is checked by crowding it cannot be corrected.

Soil for the larger rooted winter radishes should be deeply prepared and heavily manured to grow them to full size. On heavy soil grow on ridged rows or raised beds for faster drainage.

PLANTS PER PERSON/YIELD

Quick maturing salad types — Grow 40 to 50 radishes for each person. This should yield 4 to 5 bunches. Make two or three plantings a week apart to spread out the harvest. Winter types — A row 5 to 8 feet long is adequate for each person. These types develop roots 2 - 4 inches across and 8 to 10 inches long, sometimes longer. If you are growing them out of curiosity, plant only a few seeds to let you experiment with cooked or pickled radishes.

SEED GERMINATION RATE

Soil Temp.	41°F.	50°F.	59°F.
Days to Germ.	29 days	11 days	6 days

Soil Temp.	68°F.	77°F.	86°F.
Days to Germ.	4 days	3 days	3 days

Radish seeds are sturdy and dependable; they will come up under adverse conditions.

DIRECT SEEDING

Direct seed salad radishes by placing about 15 seeds per foot of row and covering ¼ inch deep. Or, seed in wide bands, scattering seeds about 1 inch apart and covering lightly. Make radish rows 15-18 inches apart. Plant winter radish seeds 2 inches apart and, since they will be seeded in late summer or fall, cover them ½ inch deep. Make winter radish rows 24-30 inches apart; the tops are robust.

TRANSPLANTING AND THINNING

Transplanting is not recommended for radishes.

Thin salad radishes twice — once when the plants are only 1 inch high and again when the roots are ½ inch in diameter and large enough to be eaten. The second time around, thin the quick maturing salad types to 1 inch apart.

Winter radishes are usually thinned to stand 4-6 inches apart on fertile soil, twice that far apart on poorer soil or for large-rooted varieties.

MULCHING AND CULTIVATION

Salad radishes are not ordinarily mulched. In warm climates you can apply a light organic mulch to late planted radishes to lower the soil temperature around seedlings growing for fall harvest and to keep the soil moisture at a uniform level.

For the large, late maturing radishes, organic mulches 4 to 6 inches deep work well. If you use fresh straw or raw sawdust, sprinkle a little nitrogen fertilizer before you lay it down. Pull the mulch up around the tops to keep them from turning rough and green. The tops of roots of certain winter varieties can protrude 4 inches out of the soil.

WATERING

Spring planted radishes do not ordinarily need watering. If no rain falls for 7 to 10 days, sprinkler irrigation to apply 1 inch of water should be sufficient.

Winter radishes are another matter, because the large roots are composed chiefly of water. If no rain falls for 10 to 14 days, apply 2 to 4 inches of water to wet the soil to a depth of 10 to 20 inches.

CONTAINER GROWING

Salad radishes are often planted in containers and harvested before warmth loving vegetables such as tomatoes are planted. They are also planted around the larger, slower maturing plants. Radishes grow rapidly in the porous, moisture retentive growing media used in containers. Canopies of clear plastic can make radish seeds sprout rapidly and bring roots into maturity early despite cold, wet weather in early spring. Plant radish seeds around summer vegetables in containers late in the growing season for fall harvest.

Winter radishes can be grown in large containers as a novelty, but are not particularly attractive. One plant is all that can be accommodated by a deep 5 gallon container.

ENVIRONMENTAL PROBLEMS

Two complaints are common with radishes: "bolting" or producing a flower stalk before the roots have reached full size, and the failure of roots to develop properly. Early planting on sandy soil or sandy loam, and thinning seedlings to one or two inches apart should cure both problems. If not, the addition of sand, organic matter and a balanced garden fertilizer are called for. Ultra-light, fast draining soils tend to dryness and their nutrients leach out quickly.

Sandy or sandy loam soils are best for radishes, as are organic soils such as muck. If you have a heavy clay soil, raise the beds for better drainage and plant radish seeds in a furrow; cover them with sand or compost. Without such precautions, clay soil could crust over the seeds and delay emergence. You may have to sprinkle late summer planted radish seeds twice daily to make them pop up in the crusted, hot soil.

HARVESTING

Under optimum conditions, half-grown roots of salad varieties can be pulled in three weeks. Begin harvesting radishes at marble size. Roots of round varieties will be good up to 1 inch in diameter, but after that will get pithy and strong flavored. Radishes should feel firm, not spongy, when squeezed.

Long rooted, white or red salad radishes take about six weeks to reach harvest size of 1 inch across and about 6 inches long. These hold well in the soil without getting strong or pithy until reaching mature bulb length.

Large round or cylindrical fall and winter radishes require 60 to 90 days and reach a diameter of three and one-half to four inches. Some tapered varieties may exceed one foot in length. Others mature at about the size of a teacup. Any winter radish can be eaten at young stages; this extends the harvest season.

STORAGE

To store radishes in a refrigerator for up to two weeks, remove the leafy tops except for one inch of stem and place in a plastic bag. Radishes are not suitable for canning, freezing, or drying. In China and Japan, radishes are pickled in brine much the same way we pickle cucumbers.

Roots of the winter radishes can be dug just before the soil freezes, trimmed to leave 2 inches of stem and covered with a deep mulch. They will keep for several weeks.

STORING LEFTOVER SEED

When placed in a sealed container with a desiccant and stored in a refrigerator, over 50% of the seed should germinate after three to five years of storage.

RHUBARB

Rhubarb

Pie Plant

Rheum rhabarbarum

Polygonaceae (Buckwheat Family)

Cool Season

Perennial

Winter Hardy In Most Climates

Full Sun Or Partial Shade

**Ease Of Growth Rating, From Crowns -- 3;
From Seeds -- 5**

DESCRIPTION

Rhubarb is a robust, long lived, large leaved perennial. Massive spreading old plants of rhubarb can reach a size of more than 4 feet high by 5 or 6 feet across. It is grown for its early crops of long, thick, greenish-red and slightly acid leafstalks, which in warm climates and under intensive sunlight tend more to green than to pink. Rhubarb is used in pies, for stewing, and in jams and wine, but only the leafstalks are edible. All parts of the rhubarb plant contain oxalic acid, a toxic substance which, if consumed in excess, can interfere with the body's utilization of calcium. However, the oxalic acid in the leafstalks is so weak that they are safe to eat. Under no circumstances should leaves or roots of rhubarb be eaten.

To produce succulent spring growth, rhubarb requires a dormant period. For this reason, and its susceptibility to crown rot fungi, rhubarb will not grow well in the low elevation areas or in rainy, humid areas of zones 5, 6, and 7; but it will thrive for years in dry, warm parts of the West above 2, 500 feet elevation.

Rhubarb is native to Siberia, China, and Tibet, and was introduced into Europe in the 17th Century. The first mention of its use in North America was in 1778. Other species of rhubarb have been in cultivation for much longer, including one mentioned in a Chinese reference as being used medicinally some 2,500 years before the Christian era.

SITE

Place your rhubarb bed to the side or in the back of the garden or at the very rear of your lot so that you will not have to walk around it whenever you wish

to harvest salad crops. Good to excellent drainage is a must with rhubarb; select a site where rainwater does not stand, even after downpours. In zones 5, 6, and 7 locate your rhubarb bed where it will receive afternoon shade; for example along the east or north side of a building or a fence. Take care to locate rhubarb where it will not have to compete with tree roots for water and nutrients. Shade is not required in zones 1 to 4; a location along the south side of a basement wall is a favored site. Rhubarb will begin to grow there 2 to 3 weeks earlier than in an exposed site because of heat reflected from the sun and radiated from the basement wall.

SOIL

Rhubarb is tolerant of acid soil and will grow well at pH levels of 5.0-6.8. Medium weight soils such as sandy loams or clay loams will produce the best crops, but heavy soils can be quite productive, though somewhat later warming than the loamy soils.

SOIL PREPARATION

As with asparagus, another long-lived perennial, care in preparing the soil for a rhubarb bed is essential. About 6 inches of manure, compost, or other organic matter should be worked in. If you have the strength, excavate to a depth of 18 to 24 inches and mix the soil and compost thoroughly as you dump it back into the trench. Try to keep the subsoil and topsoil separate and replace the amended subsoil first. Mix-in superphosphate or rock phosphate and, if you are in a rainy, acid soil area, ground limestone. Sandy soils will need liberal applications of manure or compost to hold the water and nutrients necessary to support rhubarb. The addition of substantial amounts of organic matter will raise the level of the soil 4-6 inches above the surrounding surface, and help to provide fast warmup and good drainage.

PLANTS PER PERSON/YIELD

Two rhubarb plants for each person should yield 10-12 pounds of leafstalks over a cutting period of 8-10 weeks.

SEED GERMINATION

Rhubarb seed usually germinates 60-70%. It is a minor crop, so rarely grown from seeds that germination rates at various soil temperatures have not been established.

CROWN PLANTING AND THINNING

The simplest way to start rhubarb is to use dormant divisions called "crowns." Plant them with the bud or growing point at soil level and as the leaves grow, draw up 2-4 inches more of soil mixed with compost or manure around the crowns. Don't plant the crowns deep initially because rhubarb is subject to crown rot; ignore out-dated information on planting crowns in trenches. It is important to plant clean, disease-free crowns. You should think twice about accepting crowns from a neighbor, since they will not have been grown under the same rigid disease control measures as those from a commercial grower. Crowns are available from garden centers and mail order nurseries during the late fall to early spring "bare-root season."

Thin direct seeded plants or set in crowns 36 inches apart. When thinning plants grown from seeds, remove those which are weak, pale, or of small size. Save the vigorous dark colored seedlings. Also remember that rhubarb is a long-lived perennial and should be divided every 5 to 10 years. The production of numerous small stalks is a sign the crowns are becoming crowded and need to be divided.

MULCHING AND CULTIVATION

Black plastic is an ideal mulch for new beds of rhubarb. Its use will hasten the growth of rhubarb and reduce the incidence of weeds during the critical early growth period. It will begin to deteriorate after a year or two and can be stripped off and removed at this time. Once rhubarb is established, you can switch to an organic mulch such as straw or dried grass clippings. Apply 2-3 inches of mulch once a year in zones 1, 2, and 3, right after the cessation cutting. In zones 4, 5, 6, and 7, apply mulch at the finish of cutting and again in early fall. Do not draw the mulch around the crowns in rainy areas, so as to avoid crown rot.

In very early spring it is safe to hoe or rake the soil or to mulch above rhubarb crowns and roots to kill emerging weed seedlings. This is done shallow and early, well before any leafstalks show. At other times of the year, cultivation must be done by hand to avoid injuring crowns.

WATERING

Rhubarb has a deep root system and requires a minimum of irrigation water. In most climates, no irrigation is required, but during dry spells an occasional watering with 3-6 inches of water to wet the soil to a depth of 15-30 inches will insure increased production. Make a basin around each plant and fill it with water 2 or 3 times.

CONTAINER GROWING

Rhubarb makes a strikingly beautiful and productive container plant. Start with a 20 to 30 gallon container for two crowns. Place your rhubarb container in moderate to light shade. Use fast draining growing medium such as a ready-made potting soil. Such a large amount of potting soil can be expensive. A mixture of 3 parts compost with 2 parts sand and 1 part garden soil will suffice. Fortify it with ½ cup of ground limestone per gallon of mix. Rhubarb makes heavy demands on soil for water and nutrients. Feed and water the plants frequently, at least 2 or 3 times a week. During hot, windy weather you may have to water daily. In zones 1, 2, and 3 to protect rhubarb from freezing and thawing during the winter, roll the container into a corner of an unheated garage or barn. This should increase the chances of winter survival. Give the plant a light watering monthly during the winter to keep the roots from drying.

ENVIRONMENTAL PROBLEMS

Rhubarb is a heavy feeder and does not do well on infertile, sandy soil. It is difficult to maintain a consistently high level of moisture and plant nutrients unless sandy soil is highly modified with organic matter to a depth of 2 feet prior to planting. Medium weight to heavy soil will give a better harvest if you add substantial amounts of organic matter and plant the crowns on raised beds to avoid crown rot. (See SOIL PREPARATION.) Raised beds also provide good drainage, for faster spring growth and better survival in rainy, warm areas.

The decline of rhubarb plants after 5-8 years is a concern to new gardeners, but it is a natural process. When it occurs, dig up the plants in late summer, hose them off, and use a hatchet to separate into several rooted crowns. Replant them in new beds 3 to 4 feet apart. Discard the plants if they have lumpy, knotted roots infested with nematodes.

HARVESTING

After plants have grown for two full years from crowns, rhubarb will have a 3 to 4 week harvest season. However, it will take a year longer to harvest from plants grown from seeds. Rhubarb reaches full production the third or fourth year after planting and yields for about 8 years before production begins to decline. An 8 to 10 week harvest season is normal for plants in full production. To harvest, grasp the stalk near its base, twist and pull simultaneously. Take only sturdy stalks at least 1 inch thick, 10 inches long, and topped with unfolded leaves. Leave some stalks on the plant to replenish the roots. Do not cut the stalks because of the chance of damaging the roots. In zones 6 and 7, winter to very early spring is the best time for harvesting rhubarb. In early summer, tall seed stalks will shoot up; these reduce the plants' vigor and should be snapped off as soon as they appear. Eat only the stalks of rhubarb. Discard the leaves, for they contain too much oxalic acid which is poisonous.

STORAGE

Rhubarb can be refrigerated in a perforated plastic bag for up to 2 weeks.

FORCING FOR WINTER USE

If you start with vigorous two year old crowns, rhubarb is fairly easy to force for winter use. Dig the crowns and before storing, lay them on the ground in the garden and cover with a layer of dry soil to prevent their shriveling while curing. (Remember to leave some rhubarb crowns in the garden to grow a crop for following years.) After 2 to 3 weeks of freezing weather, bring the roots inside before the soil freezes so hard that prying the roots loose would injure them. Store for as long as possible at temperatures of less than 50°F. The longer the storage, the larger the stems that will grow from the forced crowns. To begin forcing, set the crowns close together or in a box on the floor of a dark or dimly lit room. Cover with sand or soil to a depth of 2-3 inches. Moisten the area occasionally. After 6 to 9 weeks at a forcing temperature of 55-60°F stems will attain good length and thickness. In the absence of light, they will grow long, with an excellent light pink color, and the leaf blades will barely begin to unfold by harvest stage.

STORING LEFTOVER SEEDS

Rhubarb seed is seldom stored but, should you wish to do so, store as recommended. Seed should germinate at least 40% after two years.

RUTABAGA

Rutabaga (Swede, Swede Turnip, Swedish Turnip, Canadian Turnip, Turnip-Cabbage, Turnip-Rooted Cabbage)

***Brassica napus* Napobrassica Group (*B. campestris* Var. *napobrassica*)**

Cruciferae (Mustard Family)

Cool Season Biennial, Grown As An Annual

Killed By Prolonged Freezing Weather

Full Sun Or Partial Shade

Ease Of Growth Rating -- 3

DESCRIPTION

Rutabagas look like giant turnips and are grown for their large, rounded roots which can be stored for long periods at low temperatures and high humidity.

Rutabagas differ from turnips in several respects; they have large, rather smooth, grayish-green leaves; tall, leafy necks and a large taproot. The taste is distinct; some consider it sweeter than turnips. The tops are smoother and more blue-green, full of fiber and not usually eaten.

Traditionally a crop for fall harvest, rutabagas may, if early spring planted in zones 3-7, become woody and fibrous before the roots grow large enough for a significant harvest. Therefore, in these zones, seeds are planted from late summer to fall. However, in zones 1 and 2, plantings can be made from spring through June, ceasing in time for large roots to form before growth stalls in cool, late fall weather.

Periods of very cool weather, 50°F and below for a week at a time, can trigger flowering in spring planted rutabagas in zones 1 and 2 but is seldom seen except in the Canadian Plains Provinces. Premature bolting can also happen in zones 6 and 7 when rutabagas are planted for maturity during the winter. Rarely do rutabagas perform well in zones 5 through 7, regardless of the planting date, perhaps due to the lack of extended periods of cool but not freezing weather.

In "rutabaga country" certain of the white fleshed varieties are planted in two flights, two to four weeks apart. The early summer planting is grown for storage roots while the later is harvested for immediate table use. The smaller, younger white roots are milder and more tender than yellow rutabagas.

Garden cultivation of rutabagas, an Old World plant, goes far back in history. The origin of the crop is lost in antiquity. Rutabagas are closely related to the oilseed crop, Rape, which is extensively grown in northern and central Europe and in parts of Canada.

SITE

In zones 1-5, rutabagas prefer full sun. If you wish to try them in zones 6 or 7, you will be planting seeds at about the time you would collards, while the soil is still warm. The seedlings would benefit from afternoon shade, preferably from deciduous trees that loose their leaves in the fall, creating a full-sun situation late in the season. Do not plant rutabagas in the same soil two years in succession and try not to follow any cole crops or other members of the mustard family. Soil borne diseases can build up.

SOIL

Rutabagas are moderately tolerant of soil acidity and will succeed at soil pH levels of 5.5 to 6.8. They are often grown at somewhat higher levels by including ample organic matter in the soil.

Roots size up best in medium weight soil with a good content of organic matter. In heavy soil, rutabaga seeds will come up slowly and roots are often malformed. In fine, sandy, infertile soil, roots rarely develop to their full potential size.

Boron deficiencies occasionally show up on eastern soils, manifested in internal brown spots in rutabaga tissue. Home gardeners routinely correct it with light sprinklings of laundry borax in the row at seeding time. However, commercial growers with borderline boron deficiencies hesitate to apply more than 2 oz. of borax per acre! The application of any boron on arid western soils should be avoided at any time. Boron toxicity, from too much of it, is a more common problem and a difficult one to solve. If you are considering applying boron, first consult your County Agricultural Agent.

SOIL PREPARATION

Rutabagas grow rapidly on raised beds heavily fortified with organic matter. The manure or compost should be well rotted and in small particles. If you incorporate large quantities of rough organic matter, grow another crop on it first, such as spring peas or early corn, in order to give the large particles an opportunity to break down. Large, rough chunks will make it difficult to grow smooth, well-formed roots.

Clubroot is a disease of mustard family members that can be a real nuisance because the lumps on roots can impede the absorption of water and plant nutrients and cause stunted, tough, bitter roots. The casual organism doesn't like well-limed soils. A pH level of 7.2 is considered effective in controlling clubroot; work hydrated lime into the seedbed at a rate recommended by your County Agricultural Agent of contact Agriculture Canada for advice. The effect of hydrated lime will wear off after one crop, but the temporarily high pH can alter the availability of other elements such as phosphorus.

PLANTS PER PERSON/YIELD

Grow 10 to 12 plants for each person. This should yield 7 to 9 lbs.

SEED GERMINATION

Rutabaga seeds will germinate at fairly low soil temperatures around 50°F when spring planted, but will emerge slowly. Summer or early fall planted seeds will germinate in 7 to 10 days at a soil temperature of 70°F. Fresh seeds germinate strongly, often 85 percent. The Federal minimum standard is 75 percent.

DIRECT SEEDING

Rutabagas are always direct seeded. Plant 6 to 8 seeds per foot of row and cover to a depth of ¼ to ½ inch, with the heavier covering reserved for light-weight, non-crusting soils. Spring plantings should be covered only ¼ inch. Make rows 18 to 24 inches apart, depending on the fertility of your soil.

TRANSPLANTING AND THINNING

Nothing is gained by transplanting rutabagas. You can try it with very young seedlings thinned out of a row but they will probably be checked in growth enough to interfere with root formation.

Thin plants to stand 6 inches apart on fertile soil, 8 inches on poor soil.

MULCHING AND CULTIVATION

Rutabagas are ordinarily not pampered by mulching because it is a robust crop that can fend for itself. However, on sandy soils where the surface temperature tends to run high, a light application of organic mulch soon after planting will hasten the growth of seedlings. A too-deep mulch can cause rotting of roots on heavy soils.

WATERING

Rutabagas need lots of water to form large, sweet, tender roots. Normally, under eastern and central conditions, they get it. Droughts can ruin a crop. If more than 10 days pass without a rain, apply 2 to 4 inches to wet the soil to a depth of 10 to 20 inches. Also, after a spell of hot, dry, windy, weather, you may notice the plants wilting and generally looking distressed. Don't wait for 10 days to pass, give them a deep drink. Wait until evening; watering in midday can scald foliage. Foliage diseases are seldom a problem; sprinkling is okay.

CONTAINER GROWING

The crop is rather coarse and ungainly for containers. Further, the monetary value of rutabagas can't match that of "repeating" vegetables such as tomatoes and peppers that make more practical container crops. If you wish to plant rutabagas in containers as a curiosity, grow two or three per shallow 7 gallon container such as a foot tub.

HARVESTING

At harvest time you may discover that your efforts in growing rutabagas have been too successful and you may have a bushel or so of large roots, 8 inches or more in diameter, real "bragging" Swedes. The problem is, it would take a large family to eat one of these roots at a single sitting. The next year, do everything the same, but plant a few days later to get fruit the size of the average grapefruit. You can begin pulling rutabaga roots when they are tennis-ball size but early in the season they will be only passably sweet. Delay harvesting most of your crop through light frosts to improve the flavor of the roots but do not allow them to freeze.

STORAGE

To store: Cut off tops, leaving 1 to 2 inches stem stubs and cut off the tap root where it is about ½ inch in diameter. Brush off dirt; don't wash the roots.

Pack the roots in crates of moist sand or sawdust at 32-40°F. Long-term storage at 45°F or higher will cause rutabagas to sprout and become woody.

Like turnips, rutabagas give off a moderate odor in storage but not nearly as much as cabbage or chinese cabbage. Actually, at the recommended storage temperature, odors should hardly be detectable.

If you are thinking about paraffin coating rutabagas like you see them in stores, don't. The wax is applied after the roots are removed from cold storage and only temporarily keeps the roots plump.

SPINACH

Spinach (No Common Names)

Spinacia oleracea

Chenopodiaceae (Goosefoot Family)

Cool Season Annual Killed By Heavy Frosts

Full Sun Or Partial Shade

Ease Of Growth Rating -- 3

DESCRIPTION

Gardeners are in debt to Popeye for popularizing spinach. He did help to promote its genuine healthful qualities and good taste, but as a potherb. Only in recent years, and particularly with the advent of salad bars, has spinach become recognized as a premier salad vegetable.

Spinach is usually the first spring salad crop to come into production and, after a long winter of high priced greens, often of indifferent quality, the first spinach harvest gives cause for celebration. The first one or two pickings can be made by snapping off outer leaves but, as soon as the first seed stalk begins to push up, the entire plant should be harvested for fresh or frozen use. The harvest season for spinach usually spans no more than two to three weeks; after that, the plants quickly shoot up seed stalks.

Spinach plants are small and can be planted rather close together; they are hardy and can withstand temperatures of 20 degrees F. if hardened. Spinach plants can overwinter in mild climates and where the snow cover is deep.

Spinach bolts, forms a flowering stalk, with the coming of the first warm, long days in late spring. Even the new hybrids that are advertised as "bolt resistant" will succumb to the reproductive urge after a week or two of such balmy weather but, with a crop as valuable as spinach, the gaining of even a few days in harvest is worth the extra expense of improved varieties.

Spinach is native to the Eastern Mediterranean, and found its way to China in the 1st Century. The Moors introduced it to Spain and it spread across Europe by the 16th Century. Spinach arrived in America with the early settlers but, despite its good taste and nutrition, is not yet one of the more popular vegetables.

SITE

Spinach can be grown in a special raised bed in the vegetable garden, or the decorative plants can be scattered among early flowers and spring shrubs. Because spinach grows early in the spring and in the fall when the weather is likely to be cool and moist, raised beds are almost a "must" except on sandy, fast draining soils. Plant spinach where it will receive full sun. In the Deep South, early fall plantings can be shaded, but the shading should be removed as soon as the weather begins to turn cold.

SOIL

Commercial growers have discovered that spinach reacts adversely to acid soils by yellowing and, in severe cases, dying. It grows best at pH levels of 6.0 to 6.8. Spinach requires high levels of available nitrogen throughout the season to make good growth. The highest yields come from clay loam soils, but spinach will also grow well on sandy or gravelly soils if they are well supplied with organic matter.

SOIL PREPARATION

Because spinach is planted early in the spring (as well as in late summer or early fall), soil should be prepared the previous fall. With this preliminary work behind you, you can merely loosen the surface and seed spinach while lower layers of soil are still frozen. Extremely early planting does not hurt spinach seeds.

Spring crops of spinach love sandy soil with its good drainage and fast warming. If your soil is heavy, make raised beds or ridge it up in rather high rows for good drainage and maximum absorption of warmth from the sun. With raised beds, you can lay down old boards or chipped wood for walkways and have access to your spinach even when the soil is soggy from rains.

Green manure crops, fall-seeded for plowing under in the spring, are of little benefit to spring-seeded spinach because they take too long to decompose. Don't apply fresh manure to beds for spring spinach. The high ammonia level can injure seedlings.

PLANTS PER PERSON/YIELD

Grow 40 to 45 plants for each person. There is no point in splitting spring spinach plantings into relay crops, because second and later plantings will probably bolt at the same time as the first planting.

SEED GERMINATION RATE

Soil Temp.	32°F.	41°F.	50°F.	59°F.
Days to Germ.	62 days	27 days	16 days	13 days
Soil Temp.	68°F.	77°F.	86°F.	95°F.
Days to Germ.	12 days	10 days	22 days	No Germ.

Spinach seed is a good candidate for pregerminating or priming because it is ordinarily planted in cold soil.

DIRECT SEEDING

Begin planting spinach seed in the garden as early in the spring as the soil can be worked. Seed germination is slow in cold soil, but early plantings

produce superior quality because development occurs while the weather is still cool.

For spring plantings, place seeds 1 inch apart in rows and cover ¼ inch deep. Sow 15 to 18 seeds per foot in the summer or early autumn and cover ½ inch deep. In dry soils, make a 2 inch deep furrow and flood it with water. Let it soak in, plant and cover seeds as described above and lay a board over the furrow to keep the seeds cooler and to reduce evaporation. As soon as the first shoots appear remove the board.

Spinach makes a good crop for wide band seeding; plant seeds 2 inches apart in a band the width of a rake. Crops of spinach for fall harvest can be interplanted in warmth-loving summer vegetables.

TRANSPLANTING AND THINNING

Spinach can be transplanted when small, but is strongly taprooted and prefers direct seeding. The plants are not ordinarily thinned until they are large enough to eat. You will often find that a thinning in midseason will give you enough plants for a salad. If you continue to thin weekly, the remaining plants will be standing 6 to 12 inches apart when they are mature. At that stage you can remove the outer leaves instead of the entire plant until bolting begins.

MULCHING AND CULTIVATION

Spinach has a continuing need for the nitrate form of nitrogen throughout its life. You can meet this need by mulching with clear plastic which will warm up the soil in early spring and activate soil organisms and plant growth. These organisms convert nitrogen from the soil into the nitrate form which is available to plants.

Spinach planted in late summer or fall reacts positively to organic mulches which keep the soil cool and maintain a uniform level of soil moisture.

In parts of zones 6 and 7 where slugs, snails and earwigs are troublesome during fall and winter, spinach is not ordinarily mulched.

WATERING

Spinach is a shallow-rooted plant, therefore it will benefit from a continued supply of moisture, especially near harvest. It wilts or discolors quickly in dry soil. If no rain falls for 5 to 7 days, apply 1 to 2 inches of water to wet the soil to a depth of 5 to 10 inches.

CONTAINER GROWING

Spinach is an excellent plant for early spring and late season crops in containers. Although it grows quite well alone in containers, it is customarily interplanted with slower growing, later maturing vegetables in order to get a double crop from the same container. Spinach is highly decorative as well as delicious, and will be ready for harvest in the fast draining, quick warming soils of containers 7 to 10 days earlier than spinach from your garden.

ENVIRONMENTAL PROBLEMS

Premature bolting of spring plantings is a universal problem with spinach, and occurs because many gardeners do not realize that spinach can be seeded very early. It is not an exaggeration to state that spinach can be planted as early as you can brave cold weather and chilly soil to work up the soil sufficiently to plant seed. When the soil temperature passes 40 degrees you will notice signs of growth in spinach seeds, especially if you have covered them with a shallow layer of sand which traps heat.

Spring planting of spinach is a tricky proposition in many parts of the South where summer comes in rapidly on the heels of winter. In such areas, you may enjoy no more than one week's harvest before it moves from vegetative to reproductive growth. Once seed stalk buds appear you might as well harvest the entire crop for freezing because in a few days it will be too tough to eat. Bolting can occur in late-summer-planted crops during the fall and winter when long, cold periods are followed by a week or so of bright, sunny, dry weather. Usually, not all the plants in your spinach population will bolt at such a time.

HARVESTING

Cut or pick young leaves or use small plants whole as they are thinned from the row. When eight to ten leaves have developed, harvest the entire plant. Wash leaves well because they can be gritty with sand or soil.

When taking individual leaves, harvest no more than half the total at one time; leave the rest to support the growth of new leaves. Pinch off the branches near the main stem; the long leaf-stalks are tender and good to eat.

STORAGE & PRESERVATION

Fresh spinach can be stored in a moisture proof bag in the refrigerator for 5 days.

STORING LEFTOVER SEEDS

Under favorable conditions spinach seed will give at least 50% germination after 3 to 5 years of storage.

Summer Squash

Cocozelle, Crookneck, Italian Marrow, Patty Pan, Scallop, Straightneck, Vegetable Marrow, Zucchini

Cucurbita pepo Var. *melopepo* **(Bush Summer Squash)**

Cucurbitaceae (Gourd Family)

Warm Season

Annual

Killed By Light Frost

Full Sun

Ease Of Growth Rating -- 2

DESCRIPTION

Summer squash is grown for its fruits, which are eaten in the immature stage. The growing popularity in recent years is due to its productivity over a long season, ease of growth, and acceptance as a finger food and salad ingredient. Bush varieties predominate in summer squashes, but a few vining varieties are still sold.

Under the classification of summer squash come several distinct types: straightneck and crookneck, zucchini, patty pan or scallop, and vegetable marrow. Each type of summer squash has a loyal following, which is surprising, considering the minor differences of flavor in the varieties. All summer squash varieties are widely adapted and will succeed in cooler climates than will cantaloupes or cucumbers.

Summer squash is closely related to field pumpkins and acorn squash, but took a separate botanical trail so long ago that it now looks like a different vegetable. Examine the stem where it joins the fruit to see the family resemblance.

Summer squash originated in the New World and was grown by many civilizations prior to the arrival of Europeans.

SITE

Squash needs full sun all day and good soil. The foliage is subject to mildew, which becomes more serious when air circulation is impeded. Avoid planting summer squash where squash or related vine crops such as pumpkins were planted during the previous 2 years, or you will expose the plants to the risk of fusarium wilt, a soil borne disease. When garden space is limited, squash will grow well on or around a compost heap.

SOIL

Squash grows best on sandy loam soil that is well-drained, well fortified with manure or compost, and at pH levels of 5.5-6.8. Squash vines continue to grow and bear throughout a long productive season, and make heavy demands on soil fertility.

SOIL PREPARATION

Squash bushes are broad, and can be planted in ridged rows or in the center of 4″ wide built up beds. These beds can be dished somewhat to hold irrigation water. If you grow only a few plants, substitute low, flat topped mounds for regular beds.

Since heavy clay soil tends to remain too wet around the crowns of squash plants where most fruits are set, fruit can rot during wet weather. Raised beds, ridged rows, or mounded hills will improve drainage in heavy soil.

PLANTS PER PERSON/YIELD

Two to three plants per person, with regular harvesting to stimulate continuous production, should produce 40-60 fruits.

SEED GERMINATION

Squash seed germinates best in soil temperatures of 70°-95°F., but sometimes the emerging seedlings have difficulty surviving in the hotter soil ranges. At 80°F., summer squash seeds will germinate in 3-5 days. A heavy crust of clay soil can retard emergence.

PLANTING DATES

A single squash crop is grown in the summer in zones 1 and 2; in zones 3 and 4 two crops are possible; and in zones 5, 6 and 7 three crops can be grown back-to-back. The season in zone 5 is a bit short for three crops unless the fruit is grown from transplants.

Transplants are recommended in zone 1 and for earliest planting in other zones. Start seeds indoors 3-4 weeks prior to the indicated transplant time.

DIRECT SEEDING

Squash is usually started by direct seeding after all danger of frost has passed and the soil temperature has warmed to at least 60°F.

If you like to grow summer squash in a continuous, hedge like row, avoid skips by plant groups of 2 to 3 seeds 1 foot apart in rows 4 feet apart. If you prefer to plant in hills so that you can walk completely around each group of plants for watering, cultivating and harvesting, space 3 to 4 seeds in groups 4 feet apart in rows 5 feet apart. For uniform germination, plant seeds in a shallow hole the size of a saucer. Place them around the edge of the hole and cover ½ inch deep. In heavy soils, cover the seeds with sand, vermiculite or sifted compost to avoid crusting. For summer plantings use compost as it does not get as hot as sand.

TRANSPLANTING AND THINNING

Some gardeners use transplants to get the jump on the season, and find them especially helpful in areas with short summers. Start seeds indoors in peat pots 3-4 weeks prior to the indicated transplant dates. Place 2 seeds per pot and cover ½ inch deep. Grow young plants in full sun or under intense fluorescent light to prevent stretching. Harden seedlings for several days before transplanting to the garden, and be careful not to disrupt the roots when planting. Squash plants are sensitive to cold; be prepared to cover the seedlings if frost threatens. Space transplants 12 inches apart.

If you planted in hills, thin to the 2 strongest seedlings per location. If the seeds were drilled in rows, plants should stand 12 inches apart.

MULCHING AND CULTIVATION

Summer squash responds well to black plastic mulch in most zones. The mulch can more than pay for itself in increased yields and reduced weeding. In zones 1 and 2 use clear plastic mulch, which traps more heat than black plastic and brings squash crops along faster. On summer planted crops for fall harvest, use organic mulch rather than black plastic, in order to gain the insulating effect of organic litter around the young plants.

Keep down weeds around squash plants by hoeing, but hand weed close to the plant since leaf stalks are brittle.

SUPPORTING STRUCTURES

Nearly all modern varieties of summer squash form compact bushes. Only a few types are semi-vining. None require support.

However, some gardeners tie the main stem of summer squash plants to a stout stake driven at least a foot into the ground. As the older, lower leaves turn yellow, they are removed. This helps reduce the severity of powdery mildew because the plant is kept clean and air circulation is increased.

WATERING

Summer squash requires a steady supply of moisture during blossom and fruit development, but is deep rooted and can withstand 10-14 days without rain. When irrigating, apply 2-4 inches of water to wet the soil to a depth of 10-20 inches. On heavy clay soils it may be difficult for more than 2 inches to soak in because of the slow rate of penetration. Avoid sprinkler irrigation; wetting the foliage encourages powdery mildew disease. Summer squash responds particularly well to drip irrigation, but the emitter should be 8-12 inches from the crown of the plant.

CONTAINER GROWING

Bush summer squash makes an excellent plant for containers of 5 gallons or more in capacity. A good arrangement is a relatively shallow 10-gallon container in which 2 or 3 plants can be grown. Set the container on a low stand so that the bushes can billow out over the rims and gain more room to grow. Water frequently and feed plants every other week or so during the summer because the large leaves, heavy root systems, and continuous production draw heavily on the artificial growing medium.

Three bushes in a 10-gallon container should yield more than enough squash for a small family.

HARVESTING

The prime stage for harvesting summer squash is a few days past blossom stage, when the last remnants of the flower are still attached to the fruit. Try to harvest straightneck and crookneck squash when they are no longer than 5 inches, zucchini types at no more than 7 inches, and scallop types when they are no more than 3 inches across. The globe zucchini types should be no larger than an apple. Harvest while the skin can be pierced with a thumbnail and before the seeds grow large and tough. If they have grown beyond the recommended size, summer squash can still be used after scooping out the center to remove the tough seeds.

Keep summer squash harvested even if you can't use all of the produce, since maturing fruit drains energy from the plant and reduces future production.

STORAGE

Summer squash can be kept in the refrigerator in a moisture-proof wrap for 5 days.

STORING LEFTOVER SEEDS

Squash seeds are among the longest lived of any vegetable. When stored as recommended, seeds should germinate at least 50% after 5 years.

SWEET POTATO

Sweet Potato

"Kumara" In Australia And New Zealand

Ipomoea batatas

Convolvulaceae (Morning Glory Family)

Warm Season

Tropical Perennial, Grown As An Annual And Propagated Vegetatively

Killed By Light Frosts, Damaged By Prolonged Cool Weather

Full Sun

Ease Of Growth Rating -- 3

DESCRIPTION

Sweet potatoes are grown for their starchy, nutritious roots. The plant has lush, spreading vines that will cover a 6 foot wide bed by harvest time. Until recently sweet potatoes were regarded as a space wasting crop that used more area in small gardens than they were worth. Calculate that it takes 12-15 plants to produce enough storage potatoes for 1 person. With the plants set 12-18 inches apart, that translates into a row 12-22 feet long per person. That is a considerable amount of garden space, but with the price of sweet potatoes on the rise it will pay to look again at the economics of the crop.

Definitely a warm weather crop, sweet potatoes grow in any area where bermudagrass will thrive, and to about 200 miles farther north. That is roughly from the southern half of zone 3 through zone 7 east of the Rockies, and in the warm, low elevations of the West and Southwest.

Sweet potatoes are not yams. They are a distinct New World species which is not closely related to, and will not cross with, the true tropical yams of the South Pacific. "Yam" type sweet potatoes are those moist-fleshed varieties which bake soft and syrupy. The Jersey or dry-fleshed types bake dry and mealy. The term "yam" has been highly publicized by the Louisiana sweet potato industry and is the trademark for sweet potatoes of the moist-fleshed types grown in that state and sold as "Louisiana yams."

While sweet potatoes are a secondary crop in North America, they are critical to the diet of many societies of Asia and the South Pacific. The roots are high in carbohydrates and vitamins. Sweet potatoes offer an ease of culture and production per acre that few crops can match. The tips of sweet potato vines are also edible and can be included in mixed greens.

Sweet potatoes originated in Mexico, and are thought to have been taken from there to South

America and then to Polynesia centuries before the Europeans arrived. Christopher Columbus found sweet potatoes growing in the West Indies. By that time they had been in use for many years and several forms were known; yellow, white, red, and brown varieties have been described. All of these forms were introduced into Europe in the 16th Century. The Spanish explorers liked the sweet varieties, and the starchy lines of sweet potatoes were neglected for many years thereafter.

SITE

Sweet potatoes must have full sun, warm soil, and excellent drainage. The warmer the site, the better. Shade and standing water around the roots should be avoided. In tropical New Guinea, where the soil is often wet for long periods of time, sweet potatoes are grown on large piles of vegetation covered by soil. They plant sweet potato slips or cuttings on the top of the mounds to gain drainage and heat from the decaying vegetation.

SOIL

Sweet potatoes grow well in sandy or silt loam soils, but not as well in heavier soils. A sandy or silt loam soil overlaying clay subsoil is considered ideal. Sweet potatoes will tolerate very acid soil and grow best at soil pH levels of 5.0-6.8.

SOIL PREPARATION

On sandy soil, build up ridges to a height of from 4-6 inches with soil excavated from the pathways. On heavier soils on flat land, ridge up rows to 12-15 inches. This may seem extreme, but the technique was developed years ago by commercial sweet potato growers and has proved effective on heavy soils. If your soil is clay loam or clay, don't try to modify it. Instead, make high ridged rows or build frames to contain raised beds of sand fortified with at least 10% decomposed organic matter. Beds should be 8-12 inches deep. Work the soil to spade depth before installing the frame. Never double-dig soil for sweet potatoes, even heavy clay. Loosening lower layers allows the fleshy roots to penetrate too deeply and to grow extraordinarily long and large. Individual roots can weigh up to 10 pounds and would be difficult to dig without bruising or scarring. "Jumbos" are impractical and must be cured separately from potatoes of normal size. If you have sandy or silty soil underlain with clay, don't plow, till, or spade so deeply that you bring up subsoil. Relatively small amounts of clay can make otherwise loose soils bake as hard as a brick.

PLANTS PER PERSON/YIELD

Grow 12-15 plants for each person to get 15-20 pounds, more under optimum conditions which should be adequate for fresh use and for storage over several months.

PROPAGATION AND PLANTING

Sweet potatoes will not bloom and form seeds under ordinary garden culture. They are produced vegetatively by sprouting sweet potato roots, to grow rooted "slips" or tip cuttings. Virtually all sweet potato roots will sprout and form slips when placed in moist soil at 70-75°F. Occasionally, sweet potatoes purchased from grocery stores will not sprout because they have been chemically treated to inhibit growth.

Many people have sprouted sweet potatoes in a jar of water for foliage. The production of slips is somewhat similar. Eight to 10 weeks prior to the indicated planting time, select healthy and unblemished sweet potatoes and lay them flat in a warm location (70-75°F) on well drained sandy soil or potting soil. Cover until only the top half shows. Keep the soil slightly moist. Within a few weeks shoots will develop, and quickly form roots. These "slips" may be pulled from the mother sweet potato root with the fibrous roots attached to their base. However, if there is a problem with diseases and nematodes on seed potatoes in your area, it would be wiser to cut the slips off about 2 inches above the soil line for planting as unrooted cuttings. This will reduce or eliminate the chance of introducing nematodes to your garden on the roots of the slips. Potato slips are pulled or cut from the root when they are 8-12 inches tall.

In the southern portion of zone 5 and in zones 6 and 7, where sweet potatoes can be planted over an extended period, the vines can also be propagated from 1 foot long cuttings taken from the initial planting. Cuttings root readily in moist soil if set 4-6 inches deep in the row. If the soil is dry, water each plant well. Plant either rooted slips or unrooted cuttings 12-15 inches apart in rows 4-6 feet apart after all frost danger is past. Sweet potato cuttings and slips will form roots along their entire length like tomatoes. Set slips or cuttings deep enough in the soil to cover the stem up to the second set of leaves from the top.

Thin sweet potato plants to stand no closer together than 12-15 inches in rows 4-6 feet apart. Closer spacing might encourage the formation of small roots. As a rule, the 4 foot spacing is used for "bunch" varieties and the 6 foot spacing for varieties with standard, rampant vines.

MULCHING AND CULTIVATION

Good weed control is essential to high sweet potato production. Mulching can improve production and simplify weed control. Along the northern border of the range of adaptation (the line between zones 2 and 3), mulch sweet potatoes with 4 foot wide strips of clear plastic. In midsummer, when weeds begin to grow under the clear plastic, cover it with a 2-3 inches layer of organic matter to shade out the weeds.

Elsewhere, black plastic makes an excellent mulch for reducing weed problems and for increasing yields. In zones 6 and 7, mulching with straw or hay in midseason is recommended. Until that time, runners can be picked up and turned back to expose the soil for cultivation. Once the organic mulch has been applied, weeds will not come through it, but runners may strike roots if the mulch is not too deep. In some varieties the sprawling, rampant vines of sweet potatoes may set their fleshy roots only near the "mother plant." In other varieties they may also form wherever the vines strike roots — in some instances 2 or 3 feet away from the mother plant. In sweet potato culture the object is to increase the size and number of roots growing around the crown. Smaller roots are formed at points away from the crown of the mother plant.

When cultivating, pick up the runners carefully and turn them back on the bed. If you space rows 4 feet apart, hand hoe. With 6 feet spacing, you can set your tiller shallow and create a dust mulch to kill weeds. Lay the vines back in place so they do not grow and tangle around the crowns of the plants. After 1 or 2 cultivations, the vines will anchor and can't be turned back.

SUPPORTING STRUCTURES

Sweet potatoes are ordinarily allowed to sprawl, but if you are growing only a few plants in a small garden, you can train them up 4 feet high trellises or posts. This might cause some reduction in yield if the variety forms roots out on the runners as well as at the crown.

WATERING

Sweet potatoes are one of the most drought resistant vegetables, but moderate rainfall, evenly spaced throughout the growing season, is desirable. Irrigation is indicated if no rain falls for about 15 days. Drip irrigation under plastic mulch makes an ideal combination. Sprinkler irrigation is preferred to furrow because, late in the season it is difficult to get water through the dense vines. Apply 2-3 inches of water to wet the soil to a depth of 10-15 inches. Experiments have shown that sweet potatoes grown on moist soil form roots nearer the surface of the ground.

CONTAINER GROWING

Sweet potatoes grown in large, rather shallow containers will produce a fair number of good sized roots. The plants are quite decorative, with dark green or purplish vines and heart shaped leaves cascading and trailing for long distances. Each vine will need a container of about 10 gallons of sandy soil mix. Feed container grown sweet potatoes a fertilizer high in phosphate and potash every 2 weeks until the plants are about half grown; then reduce the frequency to feed every 3-4 weeks. They will need a lot of water to overcome the high rate of transpiration.

ENVIRONMENTAL PROBLEMS

In growing sweet potatoes, the object is to increase the number and size of roots growing around the crown. By nature sweet potato plants will take root wherever a joint strikes soft ground; smaller roots may be formed at these points, depending on the variety. Distorted roots can come from clay pockets in the soil or chunks of coarse organic matter, as well as from diseases or nutrient deficiencies.

Skinny or small roots can be due to late starting, to water shortages; or to extreme deficiencies of potash. A magnesium deficiency is signalled by yellow areas between leaf vines on older leaves, or by reddish purple on the Porto Rico variety, and in severe cases the older leaves will turn brown and dry. Magnesium deficiency is common on sandy acid soils. A boron deficiency shows in curled leaf stems on gnarled, stunted vines, with the older leaves dying and dropping. The roots are misshapen and rough, often lop-sided or slender, spindle or dumbell shaped. The skin may show rough areas and occasional the roots are discolored internally.

Sweet potato weevils have become such a problem in parts of the deep South that growing sweet potatoes is not worth the effort. In fact, certain infested areas are under weevil quarantine and sweet potatoes may not be shipped from there to weevil-free areas.

HARVESTING

Wait until cool autumn weather yellows the vines, but don't delay until a hard frost kills the tops. Make a test digging under 1 plant; if the roots are 2-4 inches in diameter they are ready to harvest. Cut off and remove the vines. Dig the roots, working carefully to avoid bruising. Sweet potato skins are tender and easily marred. Instead of washing the roots, brush off the soil when it is dry.

STORAGE

To keep well, sweet potatoes must first be "cured" for 10 days or more at a temperature range of 80 to 95°F and a relative humidity below 70%, all the while maintaining good ventilation. Curing will go faster when the air is relatively dry. The object in the

process is to make the roots lose about 8% of their original weight.

A curing room can be as simple as a room in a garage where a window admits sunlight for heat and a small circulating fan keeps air moving. Coldframes can also be used, with burlap laid over shallow trays of roots held off the floor by bricks. Tilting up one edge of the coldframe lid keeps the hot air flowing out and maintains ventilation.

After curing, sort the roots again. Lay aside for fresh consumption any roots that have significant cuts and bruises, and all that are pitted or discolored.

Store in slotted crates in a cool, dark place, 55-60°F, to keep roots for up to 4 months. Resist the impulse to periodically sort through and remove roots that are going soft or rotting. Research has shown that even a slight disturbance from careful sorting causes storage diseases to spread faster. Rots take several weeks to spread and even the potatoes with decayed spots can be trimmed and salvaged. Only when decay is serious should roots be sorted, the good ones dipped carefully in a 10% bleach solution, air dried and returned to storage. Such disinfection may have only a slight effect on slowing down rots, but it makes you feel that, at least, you are working to stop it.

Once sweet potatoes are brought into the kitchen, they should not be refrigerated, but kept at room temperature in a moisture proof bag for up to 5 days.

Storage will affect the flavor of sweet potatoes. Recently cured and stored potatoes will be sweeter than those that are eaten fresh after harvest. In storage, the starches in the root convert rapidly to sugars. This improves flavor, especially in the dry-fleshed varieties, and brings a softer texture. This improvement will be especially evident in the canned or processed product.

STORING ROOTS

In zones 6 and 7, it is difficult to maintain a 55-60°F storage temperature. Consequently, roots rarely keep long enough to use for slip production the following spring. The problem is different further north, where the 4-month limit on storing potatoes is shorter than the average span of winter. If you need only a few plants you can save a few tubers, root them in water, and hold them through the winter on a sunny window sill.

SWISS CHARD

Chard (Swiss Chard, Spinach Beet, Seakale Beet, Silver Beet)

Beta vulgaris Var. *Cicla*

Chenopodiaceae (Goosefoot Family)

Biennial Or Short-Lived Perennial, Usually Grown As An Annual

Winter Hardy In Most Climates

Full Sun Or Partial Shade

Ease Of Growth Rating -- 2

DESCRIPTION

Easy to grow swiss chard is related to garden beets but does not form edible roots. It produces perhaps the heaviest and most sustained crops of greens of all vegetables and makes a welcome summer treat when the spring greens have gone to seed. It is one of the few nationally adapted vegetables that will produce during most of the growing season.

Chard has erect leafy plants with tightly clustered inner leaves and arching outer leaves. Plants grow 14 to 24 inches high and up to 24 inches across. Chard is grown for its leaves and leafstalks which can be cooked together or separately. Some cookbooks recommend using chard raw in salads but, even in young stages, the raw taste is so strongly beet-like that it can dominate the other salad ingredients unless used with discretion.

When chard over-winters in the mature stage it will usually, but not always, go to seed the following spring. When and if it does, the plants are of little further use as a vegetable and should be pulled out. You can harvest the tender, young seedbuds for cooking, but it is hardly worth the effort.

Until recently, chard had been neglected by most USA plant breeders. Conscientious seed marketers have had to go to Europe for seeds of modern, attractive varieties.

Chard is native to lands around the Mediterranean and is thought to be the forerunner of today's garden beets. In 350 B.C. Aristotle described the red type and Theophrastus wrote about the light and dark green types. The yellow type, now rare, was described by a Swiss botanist in the 16th Century.

SITE

Chard will endure more shade than most other vegetables, but will grow open and lanky if the shade is too dense. If possible, give the plants full sun at least six hours each day. In zones 6 and 7, chard benefits from afternoon shade.

SOIL

Chard will thrive where more demanding vegetables would fail. It is a close relative of garden beets and, similarly, is only slightly tolerant of soil acidity. Chard grows best at soil pH levels of 6.0 and 6.8.

For sustained production of tender leaves, chard must grow quickly, and prefers moderately well drained soil such as sandy loam or clay loam. Heavier soils will produce a good crop of chard, but the major production will not come until late in the season.

SOIL PREPARATION

Work organic matter into the top 8 to 12 inches of the soil. Make sure that the particles of organic matter are fine such as in well-rotted, screened compost, peat moss or forest compost. Large chunks could interfere with germination. If your soil tends to alkalinity, with a pH higher than 7, you might incorporate agricultural sulfur, per the directions on the package, to drop the pH to about 6.5. Except on sandy, extremely fast draining soils, swiss chard grows better if you raise the beds for good drainage or plant seeds on ridged rows.

PLANTS PER PERSON/YIELD

Grow 6 to 8 plants per person. After growing until the leaves are the size of one's hand, swiss chard will produce continuously through the growing season, yielding 10 to 12 pounds of leaves from 6 to 8 plants. This yield can be increased considerably by heavy feeding and watering and by frequent harvesting. Heavy harvesting does not hurt the plants.

SEED GERMINATION

Chard seeds come in clusters within a corky coat. Two or three sprouts may emerge from a single multiple seed. Seeds usually germinate about 70% but sprout unevenly over a two to three week period.

SOIL TEMPERATURE
DAYS TO GERMINATION

Soil Temp.	41°F.	50°F.	59°F.	68°F.
Days to Germ.	42 days	19 days	10 days	6 days
Soil Temp.	77°F.	86°F.	95°F.	104°F.
Days to Germ.	5 days	5 days	5 days	No Germ.

DIRECT SEEDING

Sow seeds 2 to 3 inches apart and cover ½ inch deep with sand, vermiculite or finely sifted compost. Chard seeds germinate slowly, and sprouting can be impeded by a crust of heavy soil. Plantings can be made a week or two before the last expected frost in the spring, and late summer sowings can be made in zones 3 through 7. In zones 1 through 4, a single spring planting, if sufficiently large, can produce all the chard you need for the season. However, further south and west, spring chard tend to grow too large and woody during the summer and a second planting for fall harvest is customary.

TRANSPLANTING AND THINNING

Swiss chard can be transplanted to get an early start on the crop, but because the seeds are cheap and strong growing, most gardeners prefer direct seeding. You can start seeds indoors 6 to 8 weeks prior to the date for direct seeding for your zone. Harden the little seedlings carefully because they will be quite small when you transplant them. They can be grown in peat pots for maximum survival and minimum damage to roots at transplanting.

Transplants should be set in the garden 8 to 10 inches apart in rows spaced 18 to 24 inches apart. This spacing may seem like too much at first, but the seedlings will rapidly fill in.

When thinning direct seeded chard, don't worry about thinning the multiple sprouts which come up from seed clusters. The strongest seedling in the cluster will eventually dominate and will not be delayed in maturity. Surplus seedlings can be successfully transplanted to other spots in the garden when small.

MULCHING AND CULTIVATING

Swiss chard grows better if mulched with a deep layer of shredded leaves, straw, hay or dried grass clippings to keep the soil cool, moist and weed free. The mulch should be applied after the soil has warmed. If you prefer clean culture, hand pull the weeds in close to plants and scrape off the balance with a hoe. Chard seedlings grow slowly and can be crowded out by aggressive weeds if you are not prompt and regular with weeding.

SUPPORTING STRUCTURES

Chard plants are erect and self-supporting.

WATERING

Sprinkling 2 or 3 times daily is especially important for seed germination and early plant growth.

Later, apply water if no rain falls for 10 to 14 days. Put on 2 to 3 inches to wet soil to a depth of 10 to 15 inches.

CONTAINER GROWING

Few container plants are as foolproof as chard. Seeds can be sprouted early under clear plastic in containers, and the tubs or buckets can be moved to a protected area in the fall to keep the supply of greens coming longer. Grow three chard plants per 3 gallon container. They can be crowded with little reduction in yield. Frequent watering will be necessary because the large surface area of the leaves evaporates a lot of water.

ENVIRONMENTAL PROBLEMS

One of the major problems with swiss chard is more of a flaw than a disqualification. Some seed growers have let their lines of swiss chard go too long without renewal and off types have crept in. These off types may vary widely in color, stem characteristics and plant habit from the desired type. If your planting of chard produces more than 5% of these off types, you should complain to the seed marketer.

STORAGE

Swiss chard is highly perishable and should be kept in the refrigerator in a moisture proof bag only 1 or 2 days.

FORCING FOR WINTER USE

Swiss chard plants can be watered deeply and moved to a coldframe with a large rootball of soil. They should be set into the soil of the coldframe at approximately the level occupied in garden soil. If trimmed back, fed and watered, they will resume growth shortly. Under protection, the plants will continue to produce leaves well past the time when the top growth would be killed by frost.

STORING LEFTOVER SEEDS

If placed in a sealed container with a desiccant and stored in a refrigerator, swiss chard seed should germinate at least 50% after 5 years.

TOMATO

Tomato (Called "Love Apple" In Colonial Days)

Lycopersicon lycopersicum (Formerly Lycopersicon esculentum)

Solanaceae (Nightshade Family)

Warm Season

Perennial, Grown As An Annual

Killed By Light Frosts

Full Sun Or Partial Shade

Ease Of Growth Rating, Bush Varieties -- 3; Staked Varieties -- 4; Trellised Varieties -- 5

DESCRIPTION

Tomatoes are the most popular garden vegetable in North America. They are grown in virtually every garden . . . indeed, in some gardens, tomatoes are the only vegetable.

Garden fresh tomatoes are not only delicious and nutritious but also the most valuable garden vegetable in terms of the amount of fruit yielded times the average price per pound for the harvest season.

No other vegetable has attracted more attention from specialists, yet the beginning gardener can still find that deciding which tomato to plant, where, when and how, can be complicated. Apprehension often clouds the first season of growing tomatoes, but most gardeners are successful on their first try.

To describe the "typical tomato" one would have to ignore the extremes, yet in the extremes is where tomatoes have so much to offer. Look at the choices and, in the process, you will grasp the scope of the tomato genus:

SIZE OF FRUITS. Small cherry tomatoes are only ½" in diameter. Yet, individual tomatoes of more than 6 pounds in weight have been authenticated by the Guinness Book of Records. The average slicing tomato would weigh 5 to 8 ounces.

COLOR OF FRUITS. Tomatoes are red, but you can also buy named varieties with pink, purplish-red, orange, white, yellow, golden, or striped fruits. Breeders have other colors in the offing.

SHAPE OF FRUITS. Round, oval, squarish in cross section, elongated like a sausage, heart-shaped, pointed, flattened and turned up like an old shoe . . . take your choice.

QUALITY OF FRUITS. "Quality" in tomatoes includes a number of positive factors. The ideal home garden tomato has thick walls, firm flesh with few seeds, a pleasant, moderately acid taste free of off-flavors, thin skin with few or no cracks, uniform ripening without green shoulders, no hard core, and stems that separate easily. Quality has nothing to do with fruit size.

SIZE OF PLANTS. Tomato plants can range in size from miniature pot varieties 6 x 12 inches to large, indeterminate varieties with vines 6 x 2½ feet. The average sized midseason tomato has vines 4 to 5 feet in height by 2 feet in spread.

SEASON OF MATURITY. Days from transplanting to first harvest can vary from 50 days in the small-fruited, small-vined, extra early varieties to 80 days for the large-fruited types.

CONCENTRATION OF FRUIT SET. Some varieties, developed for canning, mature a great many fruits within a short period of time. Others bear as many or more fruit but over a longer time.

FOLIAGE. You have to look twice at certain tomato varieties which are not yet in fruit to distinguish them from potatoes. The "potato-leaved" plants have thick, rough, dark green foliage.

Armed with this information, you can be selective about the tomato lore that will be coming at you from all quarters. You will find that most experienced gardeners are utterly convinced that their way is *the way* to produce a good tomato crop. The fact is that you have all sorts of latitude in tomato varieties and culture.

When first introduced, tomatoes were regarded as a decorative plant. The natives of Peru and Ecuador ate tomatoes, and Mexican tribes grew them among crops of corn.

The first variety taken to Europe may have been a golden colored selection because early references were made to "pomi de oro" (golden apple). Adoption of tomatoes by Americans was slow because of misplaced northern European folklore about their poisonous qualities. The original European varieties were not well adapted to the U.S.A. It remained for pioneer American plant breeders around the turn of the century to select early and heat resistant varieties from survivors in disease infected fields to get what was called "field resistance to diseases".

SITE

The better the soil, the better the tomato crop. Avoid sites where water stands after a rain or irrigation, or where foot traffic is likely to compact the soil around the plants.

In zones 1 through 5, plant tomatoes in full sun. In the deep South and desert West, tomato plants are often placed where they will be shaded from afternoon sun. This reduces the stress on plants that can cause cracked or sunburned fruits. In desert areas light shade also reduces infestations of leafhoppers that carry curly top virus disease.

Rotate tomato crops; don't plant them in the same site two years in succession or after a crop of potatoes. To do so would increase the possibility of soil borne diseases.

The height of staked or caged tomato plants of a given variety bears on where you place the tomato row in your garden. Vigorous "indeterminate" varieties can grow to a height of six to eight feet. Extra early "compact determinate" varieties are usually allowed to sprawl but can be contained within a cage. Caging is usually done to conserve garden space. Short cages do not cast a significant amount of shade on adjacent rows, but shading of adjacent vegetables is certainly a factor with the taller varieties.

SOIL

Tomatoes are moderately tolerant of soil acidity and will grow best at pH ranges of 6.2 to 6.8.

Light, fast draining, "warm" soil is an advantage for growing tomatoes in zones 1 through 3 if the soil is not permitted to dry excessively. Elsewhere, tomatoes will tolerate a wider range of soil conditions. Although the early varieties will do best anywhere on sandy to sandy loam soils which warm up quickly and drain fairly rapidly, maincrop and late varieties will do best on heavier soils which will sustain growth and development over an extended period.

If your soil is medium to heavy clay, it would be difficult to modify it with sand sufficiently to grow early varieties. Instead, consider planting early varieties in containers; reserve your garden space for the larger, more rugged and productive plants of maincrop varieties.

SOIL PREPARATION

Tomatoes are accommodating. There are dozens of equally effective ways to grow them, all of which will produce tomatoes that taste the same and are of equal food value.

One proven method of growing tomatoes uses a pen made of wire fencing, 3 to 4 feet high by 4 to 5 feet in diameter. Fill it with a mixture of equal parts of decomposed compost, manure — the fresher the better — green matter and dry materials such as leaves or straw. Wet down each 6 inch layer as you build the pile. Dig the soil deeply in a 2 foot wide band around the pen and incorporate a 2-3 inch layer of decomposed compost or manure. Mix in phosphate sources to spade depth, as you prepare the soil or band it around the transplants, 2 inches deep and 6 inches away from the plants.

This setup will provide a massive reservoir for gradual release of water, and a continuing trickle-down of plant nutrients as the compost decomposes. The tomato vines can be trained and tied up and over the pen, and may even strike roots where the stems contact the moist compost.

Compost-heap tomato culture calls for large-vine varieties that can best utilize the bountiful nutrients and water within and beneath the compost heap.

In zones 5 through 7, soil temperatures run high for so long that organic matter oxidizes rapidly. This is why you rarely see naturally black soils there, except for a few pockets of peaty soil on old lake beds. Keep this in mind when growing tomatoes; the organic matter you incorporate might not be sufficient to carry the crop through the growing season. Organic mulches help to compensate.

In preparing soil for tomatoes, do not work in fresh steer or horse manure just prior to planting. It can result in excessive vine growth and poor fruit set.

PLANTS PER PERSON/YIELD

Tomato production varies widely but, in zones 3 through 7, 2 to 3 well grown plants of a standard hybrid variety will normally produce 120 to 160 4-ounce tomatoes weighing a total of 30 to 40 pounds.

A crop of this size will give you some tomatoes for canning and freezing as well as for fresh use. Further north, 15 to 20 pounds would be a reasonable expectation from 2 to 3 plants. At 50 cents per lb. that translates to $7.50 to $10.00!

SEED GERMINATION

Soil Temp.	41°F.	50°F.	59°F.	68°F.
Days to Germ.	No Germ.	43 days	14 days	8 days
Soil Temp.	77°F.	86°F.	95°F.	104°F.
Days to Germ.	6 days	6 days	9 days	No Germ.

Tomato seeds are customarily germinated indoors at a temperature of 65 to 80 degrees F. Seventy-five to eighty percent of the tomato seeds you plant should germinate under a good growing conditions.

DIRECT SEEDING

Research indicates that, in zones 3-7, excellent crops of tomatoes can be grown from seeds sown in the garden about one week prior to the frost-free date. The young seedling plants are "vernalized" by the cool air and are prompted to produce excellent crops no more than 10 days to two weeks later than the same variety grown from transplants. Plant 2 or 3 seeds in a group at each point where you wish a plant to grow, and thin to the strongest seedling.

TRANSPLANTING & THINNING

Home garden tomatoes are usually grown from transplants, either homegrown or purchased. Start seeds indoors 5 to 6 weeks prior to the indicated transplant times. Sow seeds in peat pots, peat wafers or other containers, cover 1/4 inch deep, and maintain temperatures as close as possible to 75-80°F. After germination, thin to one strong seedling per pot and grow at moderate temperatures — 60 to 65°F — in bright light to prevent the stretching which weakens the stems and produces inferior plants.

Harden transplants before shifting to the garden by setting them outdoors in an area protected from frost, intense sun and wind for 4 to 5 days prior to transplanting. After the soil temperature has reached 65°F, set tomato plants deep, with the bottom pair of leaves at soil level. Some gardeners leave only the growing point above the ground. In heavy soils, buried leaves could rot and the decay could involve the whole plant. Snip off the lower leaves and lay the stem in a trench 2 inch deep, with the growing point and a few leaves above the ground. Roots will develop along the stem.

Tomato plants are very sensitive to cold, so be sure to protect recent transplants if frost threatens.

In desert areas gardeners often find that seedlings grown in peat pots are slow to adjust to the garden and may die, particularly in unimproved sandy soil. The problem comes from the soil drying out, which makes the peat pots shrink. The tomato roots which have bridged between the pots and the soil can be sheared off. To reduce this problem, flake off the peat pot rim that stands above the level of the soil. Set the pot in deep, with at least an inch of soil covering the top. Or, use recycled containers such as 6-oz. styrafoam cups to grow tomato plants. Strip off the cup before transplanting.

Spacing of tomato plants depends on the method of training to be used. Plants that are to be staked and pruned can be grown as close together as 18 inches if you space rows 36 inches apart. Use sturdy stakes of sufficient height; some tomato varieties can grow 6 feet tall or more. If you grow tomato plants in cages or let them sprawl, provide spacing of 30 to 48 inches in all directions.

MULCHING & CULTIVATION

In zones 1 and 2, early tomato varieties planted on raised beds and mulched with clear plastic will yield heavy crops days earlier than non-mulched tomatoes. Under the clear plastic the soil will be kept moist and quite warm.

Apply the clear plastic mulch 2 weeks before planting time to make the weed seeds sprout and to warm the soil. At planting time remove the plastic, hoe out the weeds, let the soil dry out overnight and reapply the plastic the next day. Cut slits for planting. Some gardeners in zones 1 and 2 prefer to use black plastic which gives nearly as much bonus production with less work than clear plastic.

In zones 3 through 7, except for areas with very hot soil, black plastic is the preferred mulch for early tomatoes, and organic mulches for later varieties. Where soil temperatures run quite high during the summer, organic mulches are preferred for keeping the soil cool.

Even in northern areas some gardeners prefer to use organic mulches in order to keep the fruit clean and to reduce the splashing of soil particles that can cause foliar diseases. Straw or hay can be piled up to six inches deep around tall plants of tomatoes and stuffed under the branches of bush varieties to keep the fruit clean.

SUPPORTING STRUCTURES & PRUNING

Structures for tomatoes, training, and pruning are interdependent; to a large extent, one decides the other. The regional bias of many garden writers has contributed to the current state of confusion in tomato culture because they cannot understand why structures and pruning that work well in their region might be impractical in others. Tomato growth habits, pruning; humidity and its evil sister, foliage diseases; wind, and intensity of sunlight combine to shape local cultural practices. Out of these grow valid reasons for supporting and pruning certain varieties of tomatoes under certain conditions, while letting other varieties sprawl.

PURPOSES OF STRUCTURES. To hold tomato vines off the soil, thereby reducing foliar diseases, sunburn and damage from insects and other pests.

To permit closer spacing of vines which, under intensive culture, can increase the production of tomatoes per unit of garden space.

To provide a scaffold for vines, making pruning easier and faster, and giving better control over the density of vine cover, fruit size and uniformity of ripening, without sunburning.

TYPES OF STRUCTURES. Gardeners have shown great ingenuity in designing supports or adapting structures such as:

Stakes, ranging from 3 feet to 8 feet in length and driven from 1 foot to 2 feet deep to support plants of various heights. Stakes are usually at least 2 inches x 2 inches in cross section. Slender stakes tend to break during windstorms. As the main stem grows, it is tied at 6-12 inch intervals to the stake with soft ties. Metal stakes can get too hot for plants in parts of zones 5-7.

Trellises, often constructed of posts 4 inches x 4 inches x 8 feet, sunk 2 feet deep in the ground, with braces on the end posts and with heavy gauge wire runners stapled over the tops. Some gardeners run additional wires at 12 inch to 18 inch intervals and espalier tomatoes but the usual practice is to drop a single stout cord from the top wire to each tomato plant.

Smaller trellises of vertical 4 inch x 4 inch posts with 1 inch x 2 inch crossbars across the top. Strings are dropped from the crossbars for training up vines. This arrangement is best for small gardens or for container-grown plants.

Box-like frames of 1 inch x 2 inch slats, usually constructed so they will fold for storage, and made 4 feet to 5 feet in height.

Individual wire cages made of large-mesh fencing wire or the reinforcing wire used when laying concrete. A height of 4 feet is standard for determinate varieties, with 5 feet to 6 feet heights for tall, indeterminate types. Cages can be wrapped with clear polyethylene for early season frost and wind protection.

Concrete reinforcing wire can be bent into other shapes for supporting tomatoes, "A" frames for low-growing varieties or flat "spiders" with short legs to hold sprawling vines just above the ground.

Strong overhead structures such as rafter ends or a child's swing set, can support heavy duty nylon cords. After the plant becomes established, tie the twine loosely around the base of the stem and attach it firmly to the overhead structure. As the stems grow, wrap them around the twine. Suckering and single-stemming is essential with this system.

Small, patio-type tomato plants grown in containers may need the support of a short trellis or stake. Staking depends on the variety; early on you should be able to tell if it tends to grow horizontally or erectly.

Commercial growers sometimes combine stakes with nylon cords to support tomatoes. They sink strong stakes in a line every 4 feet, wrap two lengths of nylon cord around each stake, and pass them down the row, separated by the thickness of the stake. The tomato plants are threaded up through the nylon cords but are not tied to the stakes. Cords are run horizontally every 6 to 11 inches as the plants grow.

Palisades made of stiff wire fencing secured to driven stakes. The fencing forms an open circle with a gap at one side for entry. A palisade diameter of 6' to 8' will provide support for 4 to 6 large vines.

Tripods or Quadripods using 2 inch x 2 inch stakes up to 8 feet in length, depending on the mature height of the vines.

Arbors of poles lashed together with twine, similar to bean arbors. These are common where saplings are plentiful.

Ready-made cone-shaped wire supports for individual vines. Generally, these are too short and narrow for any but compact determinate varieties.

MATCHING STRUCTURES TO TOMATO VARIETIES. Overburdened, inadequate or mismatched structures for tomatoes are a common sight. Gardeners tend to underestimate the height and bulk of tomato varieties, particularly the tall, indeterminate types and to overestimate the mature size of compact determinate vine varieties.

When supported and pruned, fed and watered properly and occasionally trained, the tall, indeterminate varieties will grow to a height of 6 feet or more by the end of the season. Stakes should have an above-ground height of 6 feet. Cages usually top off at a height of 5 feet; a minor amount of billowing over the top won't hurt. The top wire of trellises for tall varieties is usually 6 feet off the ground.

Shorter stakes, frames and cages can be used for the modern determinate types, usually 4 feet maximum. Some gardeners have switched entirely to these shorter-vine tomatoes to reduce the cost and effort of supporting tomatoes and storing structures through the winter.

Compact determinate and dwarf tomato varieties are not usually supported in zones 1-4; they are allowed to sprawl. However, in zones 5-7, where foliar diseases are especially severe, all types of tomatoes are grown up supports. The ready-made galvanized wire cones work well for supporting the tomato midgets.

Under certain climate conditions, larger varieties are allowed to sprawl. Gardens that are plagued with strong, drying summer winds and those where cold night air inhibits pollination and fruit set are candidates. If winds tatter your tomatoes and parch the leaf margins, try letting the vines ramble. If, even on early varieties, fruit set is delayed by cold night air, consider letting vines sprawl to place the blossoms down in the layer warmed by radiating heat from the soil.

PRUNING. You will need to decide which method of pruning you will follow, before you set plants in the garden. Severely pruned varieties can be set closer together than those allowed to sprawl. Pruning is accomplished by pinching off the "suckers" that form where leaf stems join the main stem and, to a lesser extent, by nipping off vines to keep them in bounds. Pruning is usually done weekly, and for two major purposes:

To open the foliage canopy which permits earlier and more uniform ripening, without sunburning, faster drying of foliage after rain or sprinkler irrigation, and easier access for spraying.

To concentrate the energy of plants into the formation of larger fruit, through removing excessive foliage and fruit. (This also has the effect of reducing the total number of fruits but the remainder will be earlier ripening and larger).

MATCHING PRUNING TO THE PLANT HABIT AND SUPPORTING STRUCTURE.

Tall, indeterminate varieties, when run up trellis strings, are always pruned to one central stem. Each time you see a growing point of 6-8 inches waving about, twist it around the string to keep it growing vertically. When you train these tall varieties up stakes, prune to 2 stems in zones 1-4 and 2 to 3 in zones 5-7. Double or triple-stemming gives slightly larger yields than single-stemming. When caged, tall varieties can be pruned to 2 to 4 verticals, depending on the diameter of the cage.

Medium-height, determinate varieties are rarely trellised and are usually staked, caged or grown in slat frames. Pruning can be to the same number of verticals as for the taller varieties. Cages of 2 foot diameter can accommodate as many as 4 vertical stems but the inside of the cage gets crowded by late season, complicating spraying. Watch out for the husky basal suckers that form late in the season; nip them off when they are still small.

Compact determinate and dwarf tomato varieties are seldom pruned unless they are enclosed in small cages or frames to crowd more plants into a small garden. Radical surgery can hurt these small plants: pinch off suckers you feel must be removed when they are quite small.

FERTILIZING

Many different fertilizer programs have been advanced for tomatoes. Most have been successful, not so much due to their accuracy, but more to the easy-to-please nature of tomatoes. An ideal feeding program combines early harvests with sustained, heavy production of well-shaped fruit.

There is little basis for fact in the claim that what you feed tomatoes affects the taste of the fruit. The major taste determinants are the variety grown, the stage of maturity when harvested, and the weather prevailing during the ripening period. You can grow delicious tomatoes hydroponically or, at the other extreme, in a barrel full of pulverized dry leaves fortified with sand, lime, wood ashes and rock phosphate, and fed with manure tea.

Tomatoes are moderately heavy feeders. When the soil is prepared as directed, there is no need for high-nitrogen fertilizers. In fact, they can cause too much vegetative growth at the expense of fruit. Unfortunately, there are no concentrated, fast-acting organic fertilizers naturally high in phosphate and potash, and low in nitrogen. Foliage color in non-bearing juvenile plants and fruit set in older tomatoes should also be considered. As long as the foliage color is a rich medium green, and the plants are

growing steadily or setting fruit, hold off on feeding. Just because your plants are not deep, almost black-green like your neighbors is no reason for you to fertilize. He or she may not have ripe tomatoes until you have already harvested one or two crops.

You may notice the size of tomato fruit declining late in the season, particularly with the older open pollinated varieties which don't have as much vigor as the hybrids. This is a signal that something has gone wrong with your total program of feeding, watering, insect and disease control, and pruning. Late season feeding may bring a slight increase in average fruit size but might also delay ripening of fruit.

Keeping tomato plants flourishing without overstimulating them with nitrogen is difficult on sandy or gravelly soils, especially those without a clay base. Feeding every 2 to 3 weeks may be required when the plants are in full production and perhaps after every rain or deep watering as well.

In parts of zones 6 & 7 with heavy soil, it is standard practice among commercial tomato growers to incorporate granular 5-10-10 fertilizer when spading, or to apply it in a band 2 to 3 inches deep just beyond the roots of transplants. Applied at a rate of 2 to 3 pounds per 100 sq. ft., this single feeding should suffice for the entire growing season except during very rainy years.

WATERING

A uniform supply of soil moisture is necessary throughout the growing season for maximum production and freedom from blossom-end rot. Although tomato vines are relatively drought-tolerant, a loss of production will be sustained if more than 7 to 10 days pass without rain or irrigation. Apply 2 to 4 inches of irrigation water to wet the soil to a depth of 10 to 20 inches. Furrow or drip irrigation is preferred; overhead watering can foster the spread of diseases.

One-gallon plastic milk jugs make inexpensive and effective water reservoirs for tomatoes when buried between plants with only the necks protruding. Punch holes in the bottoms so water will drip out slowly. Keep the bottles capped between watering to exclude soil, trash and insects. Water-soluble fertilizer can be measured into the jugs when they are filled.

Significant fluctuations in soil moisture may result in blossom-end rot, a physiological disorder. This is usually associated with a calcium deficiency and can vary widely between varieties. Organic mulches work to level out the fluctuations in soil moisture as well as to decrease the frequency of required watering.

CONTAINER GROWING

Tomatoes are one of the best vegetables for growing in containers, and the container size depends largely upon the mature plant size. Pots of 2 to 3 gallon capacity can be used for dwarf or extremely compact determinate varieties; 5-gallon containers for compact determinate varieties, 10-gallon for large determinate and 20-gallon and up for indeterminate varieties. Two plants of vigorous indeterminate varieties can be grown in a 32-gallon plastic garbage can. Considerably smaller containers can be used but, when the plants are loaded with fruit, the frequent watering required makes them impractical. Most families can water containers twice during warm weather — before leaving for work and after returning. Overburdened containers might need watering in midday when no one is home.

With container culture, blossom end rot can become a serious problem unless adequate available calcium is supplied. To your artificial soil medium, be it purchased or homemade, add ground dolomitic limestone at the rate of about 1/4 to 1/2 pound of limestone per cubic foot of moist (approximately 7½ gallons) medium. Incorporate it thoroughly before filling the container. This initial application should supply sufficient calcium for about four months. After that time an equal quantity should be topdressed on the container and worked in to a depth of about 2 inches. The second application is necessary only in long-season areas.

When grown in containers, only the vigorous determinate or indeterminate varieties need to be supported; more compact kinds can cascade down over the rims. A simple and unobtrusive support can be provided by attaching a vertical pipe to the side of the platform on which the container sits. The pipe should be at least six feet in height, and should have a cross pipe at the top, extending over the center of the container. From this should be dropped a heavy stout length of nylon twine attached to a stake driven by the tomato plant or plants. Prune the plants to one central stem by removing the "suckers" which grow in plant axils, where the leaf stems join the main stem. As the plants grow, twine the growing tips around the cord. In windy areas, use double or quadruple thicknesses of twine to prevent disasters.

ENVIRONMENTAL PROBLEMS

With tomatoes come problems often related to the wrong choice of varieties and to feast-or-famine feeding. These two factors can cause poor fruit set and late crops.

Some modern adapted varieties will set fruits at temperatures of 60-62 degrees F. as opposed to the 65 degree or higher threshold of older, late types. Many modern varieties will also continue to set fruits during hot spells until temperatures soar into the nineties.

Inviting an old variety, such as *BEEFSTEAK* or *PONDEROSA* into your garden sets the stage for low yields and erratic performance unless you are a whiz at tomato growing.

You penalize yourself when you purchase large, blooming or fruiting tomato plants for late spring planting. These large plants suffer so much from transplanting shock and will yield only a light crop of useable fruits. Conversely, if you start with smaller plants with six leaves and set them out early under protection, you will get the maximum crop of usable tomatoes.

Blossom end rot, which is associated with uneven soil moisture and deficiencies of available calcium, has been mentioned previously.

"Catfacing" is a distinct type of cracking which occurs around the blossom-end of certain varieties when they are subjected to weather stresses. When pollination occurs during cold, wet weather, "catfacing" can be expected. If this condition bothers your tomatoes, change to a variety resistant to catfacing.

Leafroll - Rolling of tomato leaves along the longitudinal axis is common and apparently has no adverse affect on yield. It is most apt to occur when tomatoes are subjected to cold weather when about half grown. Such harmless leafrolling should not be confused with that which occurs due to the feeding of insects on the underside of the leaves or to severe infections of mosaic virus.

Sunscald - Most prevalent on early varieties in zones 4 through 7 when they are allowed to sprawl. When near maturity, the heavy fruits pull the center of the plant open and expose the fruits to intense sunlight. Uncontrolled foliar diseases can have the same effect. Sunscald is less likely to occur if early varieties are staked or caged. Sunscald can also occur on trellised tomatoes when pruning is delayed and an excessive amount of shading from suckers is removed at one time. Sunscald is rarely seen in zones 1 & 2.

Walnut Wilt - Many gardeners have discovered that walnut trees and tomatoes do not mix. Walnut tree roots secrete a toxin called "juglone" which can damage or kill other plants and is especially lethal to tomatoes. Walnut leaves should not be used in compost that is to be applied to vegetable gardens.

Gardeners in windy areas sometimes experience damage to staked tomatoes. This can be reduced by training vines on chicken wire held a foot off the ground by pegs. Close to the ground, the vines will suffer less from wind and cool temperatures.

HARVESTING

Tomatoes taste best and have the highest vitamin content at a uniformly ripe, but not soft, stage. Peak ripeness in red, pink and orange varieties, usually comes to about five to eight days after the first color shows. Rather than carelessly yanking tomatoes from the plant, gently twist off the fruit, taking care not to break the rather brittle stems. When stacking or piling tomatoes in a basket or on a shelf, remove the stem and green caylx in order to eliminate puncture wounds.

You may want to harvest some tomatoes early, at the green stage for frying or pickling. Ripe tomatoes can be processed for juice, sauces, or catsup. For the sake of safety avoid soft, overripe fruits or tomatoes with rotten spots. They could carry toxins or organisms that cause canned tomatoes to spoil.

If you have green fruits on your plants in the fall when frost is approaching, you can pick the tomatoes or pull the entire plant. In either case, put the fruit indoors in a cool place and it will ripen with a fairly good flavor.

STORING LEFTOVER SEEDS

When tomato seeds are sealed in an airtight container along with a desiccant, and stored in a refrigerator, they should germinate at least 50% after 2 or 3 years.

STORAGE

Ripe tomatoes may be kept in the refrigerator in the vegetable crisper or in a moisture proof plastic bag 3 to 4 days.

TURNIPS

Turnip And Turnip Greens

Brassica rapa **Rapifera Group**

Cruciferae (Mustard Family)

Cool Season Biennial Grown As An Annual

Winter Hardy In Most Climates

Full Sun Or Partial Shade

Ease Of Growth Rating, For Greens -- 2; For Greens And/Or Roots -- 3

DESCRIPTION

Few vegetables are more widely adapted. Turnips grow quickly from seeds to form erect plants 12 to 18 inches in height. At near maturity the entire plant can be harvested for leaves (greens) or allowed to

151

grow edible roots. Turnip greens are often mixed with milder potherbs such as mustard greens, because they have a pronounced flavor that is heavy to some palates.

Turnip roots will store for only short periods of time in root cellars and should not be kept for more than a week at a time in a refrigerator as they can throw off a penetrating odor.

Roots of turnips may be white, purple-topped, golden, scarlet or woody, depending on the variety. There is only a slight difference in the taste. Some varieties are grown only for greens and form fibrous, inedible roots.

The aim in turnip growing is to have plants develop so fast that the leaves will be tender at large stages. Roots should form quickly and not become woody or "hot" in taste.

Turnips have been cultivated since ancient times. Seeds were brought to America with the first settlers in Virginia. Turnips are native to the Caucasus and the Baltic Sea areas and were thought to be cultivated by the Germans and the Celts. In A.D. 42 Columella wrote that the turnip was said to be good for man and beast. Distinct types of turnips are enjoyed and cultivated, some reaching 40 pounds or more in weight. These are the types favored for cattle feeding in former years.

SITE

Spring planted turnips like sandy soils or silt loams which offer good drainage and fast warmup. Summer planted turnips prefer heavier soils that maintain a more uniform level of soil moisture and available nutrients. If you have a choice, locate your turnip rows accordingly.

Fall turnips will remain in the ground late in the year when shadows are long; don't locate your turnip row near buildings or walls.

SOIL

Soil pH ranges of 5.5 to 6.8 will produce good turnips. Acid soils will require liming. Although turnips are not exacting as to soil, the roots will size up better in deep, rich loam. Heavy soils are apt to crust, especially when seeds are planted in late summer.

Turnips require more phosphorus than other vegetables and are used by scientists as indicator plants for phosphorus deficiencies.

SOIL PREPARATION

Well rotted manure is often incorporated in large quantities and pays off in production. Fresh or strawy manure or stable litter should be composted before use. Turnips can get by with rough, hasty soil

preparation when they follow another crop. A quick loosening with a spading fork and a leveling with a rake to remove debris and old roots will pulverize the surface layer for a seedbed. Work in preplant fertilizer and a thin layer of compost to keep the surface porous.

In zones 2 and 3 turnip seeds can be spring planted very early. Preparing the soil the previous fall, except for raking, will simplify planting in cold, disagreeable weather.

Build raised beds on heavy soil in western or southern areas where heavy winter rains may come while turnips are still in the ground.

PLANTS PER PERSON/YIELD

Grow 25 to 30 plants for each person to get 8 to 10 pounds of turnip roots plus a few pickings of tops for greens. For spring greens, plant 30 to 50 plants per person.

Plantings for fall or winter can be more generous because garden space is usually less restricted and harvesting is extended over a longer period. Large turnip patches are common in zones 6 and 7 where greens are frozen in quantity.

SEED GERMINATION RATE

Soil Temp.	41°F.	50°F.	59°F.	68°F.
Days to Germ.	No Germ.	5 days	3 days	2 days

Soil Temp.	77°F.	86°F.	95°F.	104°F.
Days to Germ.	1 day	1 day	1 day	2 days

DIRECT SEEDING

Turnip seed is quick germinating and grows fast. The plants reach harvest size in less than 60 days. Direct-seed in rows by placing 12 to 15 seeds per foot, or sow in wide bands with seeds spaced about 1 inch apart in all directions. Cover spring sowings ¼ inch deep, and summer sowings ½ inch deep.

For summer plantings in dry soil gardeners can open a wide furrow, flood it with water, and scatter seeds when the water has soaked in. They can then cover the seeds with sand and the furrow with a board. In the dark, moist environment the seeds will sprout quickly despite heat that bakes the surrounding soil. Remove the board gradually to let the seedlings adjust to the wind and sun.

TRANSPLANTING AND THINNING

Turnips are not ordinarily transplanted.

Thin to 3 inches between plants in rows and space rows 12 inches apart. Thin wide band plantings to 4 inches in all directions. Use the larger thinnings for salads and greens.

MULCHING AND CULTIVATION

Large patches of turnips are not usually mulched, but summer and fall planted turnips would benefit by having a layer of straw pulled up around them to help keep the soil cool and moist. Mulches also help to keep the foliage clean and the upper third of the roots smooth.

WATERING

A consistently moist soil is necessary to the formation of well formed roots and large, tender leaves. Established turnips are drought resistant, but prolonged dry weather will seriously affect harvest quality. If no rain falls for 10 to 14 days, apply 2 to 4 inches of irrigation. Sprinkler irrigation is easy and effective.

CONTAINER GROWING

Turnips make a good but not especially attractive container crop, and can be planted around summer vegetables as they begin to decline. You can cut off and remove plants of eggplant, pepper and tomato when they have been killed by frost, and the turnips will continue to grow late in the season. You can move containers of turnips under shelter when freezes are predicted, and set them out in the sun when warmer weather has returned. In this way, you can continue to harvest greens and roots until late winter.

ENVIRONMENTAL PROBLEMS

Problems with turnips can occur when gardeners grow them as a spring crop in zones other than 1 through 3. Spring planting can be successful where cool weather is prolonged, but this is limited to only a few states. In most climates, spring planted turnips are just beginning to make roots when summer arrives. The hot weather makes the roots woody and strong, and leaves become stringy and bitter. Bolting occurs rapidly.

If you have had problems with stunted or spindle shaped roots on turnips planted for fall harvest, incorporate a 2 inch layer of well decomposed manure or screened compost to a depth of 1 foot, along with a phosphate source, sow seeds a little earlier and thin plants to stand 4 to 6 inches apart. This should correct a common condition of soil being too poor and too dry, and stands of turnips too crowded for turnips to grow fast.

Avoid "checks" in growth; turnips should grow steadily and rapidly.

HARVESTING

Individual leaves can be pulled for greens any time they are from 4 to 12 inches in length. Smaller leaves are good to eat but a pain to wash and prepare. The larger the leaves the easier they are to wash. Tattered or weatherbeaten leaves are usually left on the plants. You can begin to pull roots when they are 2 to 3 inches across. Don't let them grow to the large sizes often seen in grocery stores because they may be woody, strong-flavored or bitter.

STORAGE

After harvest, wash turnips if necessary, but be sure the moisture is evaporated before storage. Cut tops leaving ½ inch above the crown. Put in moisture proof plastic bag and store at 32-40°F. Turnips give off an odor when stored, therefore, do not store in the basement. To store turnips in the refrigerator, remove leafy tops, wrap in moisture proof plastic and store in the refrigerator up to 2 weeks. To store young tender turnip greens, put in moisture proof bag in refrigerator and store 3 to 5 days.

FORCING FOR WINTER USE

Turnip greens are quite hardy to cold, and with a minimum of shelter will continue to grow well into the winter in cold climates. Turnips can also be sown very early under tightly constructed cloches to produce extremely early spring greens.

STORING LEFTOVER SEEDS

Turnip seeds are strong and fairly long lived. When placed in a sealed container with a desiccant and stored in a refrigerator, over 50% of the seeds should germinate after 3 to 5 years.

WATERMELON

Watermelon

Citrullus lanatus (C. vulgaris)

Cucurbitaceae (Gourd Family)

Warm Season Annual

Killed By Light Frosts

Full Sun

Ease Of Growth Rating, Bush Varieties -- 4;
Vining Varieties -- 5

DESCRIPTION

Watermelons range in size from 2 lb. icebox types to giants of 200 lbs. or more. The 1982 world's record melon weighed 219 lbs. and was 34½ inches in length. Most garden varieties are in the more practical sizes of 2 to 20 lbs.

Melons ripen in early summer in the deep South and low elevation desert climates of the Southwest, but not until late summer in the North and highland areas. For good growth and flavor, watermelons need daytime temperatures of 70 to 80°F and night temperatures of 65 to 70°F.

With every passing year plant breeders are making watermelons earlier, more productive and more space-efficient. Considered at one time a crop more appropriate for farms than for gardens, watermelons are now practical even in fairly small gardens.

Standard watermelon varieties are notorious ramblers and need lots of room to spread. Compact varieties are available. They are sometimes called "Bush" watermelons but their plants are prostrate and sprawling, not upright and bushy. The vines of compact varieties spread to cover a 6 foot circle but, to date, the bush melons can't match the more vigorous varieties for production and quality.

Some varieties bred for home gardens have such succulent flesh and tender skin that, when ripe, they will split under the pressure of one's hand. Varieties are now available that will mature in 65 days from seeds. This means that melons can be ripened in all except short season northern, highland or cool coastal gardens.

A mystery surrounds the introduction of watermelons to North America. While it is thought that watermelons originated in the Kalahari district of South Africa, the first French explorers reportedly found American Indians growing watermelons in the Mississippi Valley. The watermelon was reported as being grown in North Africa in the 16th Century, in Egypt prior to that, and it is thought to be one of the fruits the Israelites regretted not having during their exile in the desert.

Some races of people grow watermelons for their seeds as well as for their flesh. The seeds supply valuable oil and protein.

SITE

When choosing a site for watermelons, think big and think warm. Leave a row 6 feet across for bush varieties and 9 to 10 feet across for standard varieties, in full sun, not shaded at any time of day. In northern areas, a southern exposure on a southward-tilting slope can add a few degrees to the average soil temperature.

If and when watermelon vines grow out of bounds, they can be turned back over the foliage canopy like you would turn back the corner of a bedspread. In areas of extremely high temperatures and high light intensity such as prevail in much of the Southwest, vines can be turned back similarly to protect watermelon fruits from sunburn.

If theft is expected to be a problem, hide your melon patch inside a screen of corn plants.

SOIL

Watermelons will grow at soil pH levels of 5.0 to 6.8 but experience increased physiological and disease problems at the upper and lower extremes. Commercial growers try to keep their fields between 5.5 and 6.5. Low pH can cause deficiencies of available calcium which can lead to blossom end rot.

The sandier the soil, the better melons will grow. An ideal situation is a sandy soil overlaying a clay or clay loam subsoil within reach of root systems.

Moderately acid soils need not be limed to grow good watermelons. If you work in quantities of steamed manure or compost to serve as a reservoir of soil moisture and plant nutrients, even deep sand will grow good melons. Don't use fresh or rotted manure; it can carry fusarium wilt.

On sandy soils, melons can be planted "on the flat", but planting on low mounded "hills" or on ridges will give somewhat earlier maturity. On clay soils melons grow best in frames to which sandy, organic soil has been added to a depth of 6 to 8 inches. Make these beds 4 feet wide for bush varieties, 8 feet wide for standards, to employ standard lengths of lumber. Vagrant vines can be turned back into the enclosures.

Melons develop slowly when grown on clay loam or clay soils and, in cool summers, can grow so slowly that getting sweet, ripe melons can be chancy. The aim is to have melons "sugar up" before nights begin to turn cool.

PLANTS PER PERSON/YIELD

Grow 2 to 3 plants for each person to get 6 to 9 melons. If you have room and a sufficiently long growing season, grow more than one variety.

SEED GERMINATION

Soil Temp.	68°F.	77°F.	86°F.	95°F.
Days to Germ.	12 days	5 days	4 days	3 days

Watermelon seeds are large and easy to germinate when soil temperature is warmer than 70°F.

Some authorities caution against soaking watermelon seeds prior to planting. If any of the seeds are infected with anthracnose or gummy stem blights, soaking can spread the disease throughout the lot.

DIRECT SEEDING

Plant watermelon seeds in the garden after all danger of frost is passed and the soil has warmed to at least 65°F. Dig a shallow saucer-size hole and place 4 to 5 seeds around the edge; cover them ½ inch deep. Be prepared to protect earliest plantings with hot caps or bottomless plastic jugs.

An "old timey" way to grow watermelons in sandy soil is to dig a hole two feet deep and two feet wide,

and to mix a bushel of rotted manure into the excavated soil before returning it to the hole. When the hole is half full, ½ pound of 10-10-10 fertilizer is spaded in, or 1 pound of an organic plant food such as cottonseed meal or trash fish. After filling the hole and mounding the soil into a "hill", 3 to 4 seeds are planted. If you try this method, use packaged pasteurized manure. When planting watermelon seeds in rows, allow 36 inches between rows of bush varieties and 5 to 6 feet between rows of standard varieties. Plant seeds 6 to 12 inches apart in rows.

TRANSPLANTING AND THINNING

For a head start, especially helpful in short season areas, sow seeds indoors 4 weeks prior to the indicated direct seeding times. Place 2 to 3 seeds in each peat pot, cover ½ inch deep and keep as close as possible to 75°F for germination. Thin to the two strongest seedlings, grow in full sun or under fluorescent lights, and harden young plants for several days before planting in the garden. (See Mulching for transplanting through plastic mulches.)

Start seedless varieties indoors and keep as close as possible to 85°F until germination occurs, then lower temperatures to the mid to upper 60's.

Delay transplanting until one week past the last frost date, and be prepared to protect young plants with bottomless plastic jugs, cloche or similar devices. Protecting plants offers some shelter from bugs as well as from frost. A pinch of insect repellent inside the jar should keep away the beetles that spread bacterial wilt and that are especially damaging to young plants.

Thin seedlings to the two strongest at each location in hills. When grown in rows, plants can stand 1 to 2 feet apart without diminution of yield.

MULCHING AND CULTIVATION

Try clear plastic over raised beds if you are on the northern border of watermelon adaptability. Lay the plastic down early to trap the solar heat and to shed rain. At planting time roll up the plastic, scrape off the weeds, replace the plastic and plant well-grown melon transplants through slits. Cover the transplants temporarily with bottomless one gallon plastic bottles.

Elsewhere substitute black plastic. You will get ripe melons 7-10 days sooner than on bare ground, and they should be large and well formed.

155

If organic mulches are used, they should not be applied until the soil is warm to the touch.

Once melon vines have begun to run or "flop", it is difficult to weed or cultivate. When they start to run vigorously, scrape off the weed seedlings within the area you expect to be covered by vines.

Using a standard large-vine variety, you can produce county fair competition size melons by removing some of the small fruits before they enlarge beyond baseball size. The surest way of producing a large melon is by removing all but one fruit from each plant, but this drastically reduces the total number of fruits harvested.

Another approach to larger fruit involves the early removal of the fruits that are misshapen or otherwise may not develop normally. The remainder of the early set is allowed to develop, and all later-setting fruits are removed. None of the vine growth or leaf area should be removed.

SUPPORTING STRUCTURES

The fruits of watermelons are heavy and the vines are allowed to run on the ground instead of supports. If you have only a few vines you can, with some difficulty, put slings under each fruit, but the accumulated weight of the fruits could tear down weak structures.

WATERING

Watermelons must have a steady and abundant supply of moisture up until the largest fruits have grown to 75 percent of their potential size. Research indicates that the sugar content of melons will be higher and the flavor better when they are grown at moderately low soil moisture levels during the ripening process.

If no rain falls for 7 to 10 days, apply 2 to 4 inches of water to wet the soil to a depth of 10 to 20 inches. Once vines have begun to run it is difficult to water by furrow irrigation. A good arrangement, in order to avoid wetting the foliage and encouraging diseases, is to turn a soaker hose upside down along the row when the plants are still small. When you wish to water, connect the hose and let it trickle for an hour or two.

Watermelons can be grown in containers, but coping with the long vines can be troublesome. The compact vined varieties are more practical for growing in containers of 10-15 gallon capacity. An advantage to growing watermelons in containers is that they can be placed on a concrete patio or against a west-facing wall to gather reflected heat which will bring them to maturity faster. Set the container on a low stool or inverted container for good drainage.

ENVIRONMENTAL PROBLEMS

Poor pollination can cause poor fruit set and misshapen melons. Insects, principally bees, are needed for pollination, and these can be scarce in neighborhoods where indiscriminate spraying is common. You can "make like a bee" by using a camel's hair brush to transfer pollen to female blossoms, which are recognizable by the swelling behind the flower. The female blossoms are most receptive immediately after they open. You will know that your efforts have been successful if, within a day or two, the stem behind the ovary begins to lengthen and the baby fruit tips down. Interplanting bee plants such as monarda (bee-balm) or borage may draw bees. Failure of fruit to ripen is most often due to the choice of a variety that is too late for your area, or the failure to plant soon enough.

Blossom-end rot can be a serious problem with watermelons. It is caused by deficiencies of available calcium and wide fluctuations in soil moisture levels. If a soil test has proven your soil to be acid, incorporate sufficient ground dolomitic limestone to raise the pH to between 6.0 and 6.5. Don't use the acid forming sulphate-based, nitrogen fertilizers. Instead, use fertilizers that tend to raise soil pH slightly, such as potassium nitrate, calcium nitrate or sodium nitrate. If blossom end rot persists after you make these adjustments, raise the beds and add organic matter for improved drainage. This should compensate for surplus moisture from rains or overly zealous irrigation.

Blossom end rot can be especially troublesome on container grown watermelons because the calcium added to the growing medium can run out at about the time fruits are ripening. To avoid it, scratch limestone into the top 2 inches of the soil at the rate of about 2 to 3 tablespoons per gallon of container capacity. Powdered gypsum, calcium sulphate, will work faster and will provide some sulphur as well.

Some gardeners worry that other vine crops might cross pollinate their watermelons, influencing their flavor. Not so. Only citron can cross with watermelons and the effect of the cross doesn't show until the next generation. In other words, you would have to save and plant the seeds from the mixed up melons to detect a difference in flavor, color and fruit size.

HARVESTING

Watermelons have three indicators of ripeness, and the most reliable method depends on the variety. You may have to check ripeness by all three to be sure.

First, check the underside or "belly" where the fruit touches the ground. The color changes from white to yellowish when the melon is ripe. Then, look for a slender, curly tendril on the stem opposite where the melon is attached. When this tendril dries and withers, the melon is ripe. Next, thump the fruit with your knuckles. A sharp "plink" or ring indicates undermaturity. A dull, hollow "thunk" should indicate ripeness.

Lastly, despite its barbarity, you can "plug" melons to be absolutely sure. If the melon isn't fully ripe, replace the plug and check it later. Rarely does plugging cause rotting if you make the incision on the side of the melon where it will not catch rain.

STORAGE

Will keep 2 to 3 weeks at 40-50°F. At 32°-40°F, watermelons are subject to chilling injury, and tend to become pitted.

STORING LEFTOVER SEEDS

Watermelon seeds are long lived. When placed in a sealed container with a desiccant and stored in a refrigerator, at least 50% of the seeds can be expected to germinate after 3 to 5 years.

WINTER SQUASH

Squash, Winter Storage

Cucurbita maxima **Winter Squash**

Cucurbita pepo **Var.** *pepo* **Acorn Squash**

Cucurbita moschata **Winter Crookneck Squash**

Warm Season Annuals, Killed By Light Frost

Full Sun

Ease Of Growth Rating, Bush Varieties -- 3; Vining Varieties -- 4

DESCRIPTION

Winter squash is grown principally for its hard-shelled mature fruits which can be kept in dry storage and used for baking and pies during the winter and early spring. The fruits can also be eaten like summer squash at very small sizes.

Major classes of winter storage squash include: Acorn, Banana, Buttercup, Butternut, Delicious and Hubbard.

Winter squash requires considerably longer to mature than summer varieties. It is more popular in zones 1 through 4 than in the South and West. In the South and warm West, spring planted squash matures in midsummer when the taste is bland due to the heat. But, plantings made after midsummer in the South are subject to severe insect and disease problems; thus, winter squash has never become an important crop in the South. Sweet potatoes displace them in many areas. With the sweet potato weevil complicating potato production in many southern areas, winter squash may gain acceptance.

Plant breeders have made great strides in developing "bush" varieties of winter squash. True, the bushes are large, four or five feet across by up to two or three feet tall; but they don't ramble all over the garden like the vigorous vining varieties which can spread to cover a 20 foot circle.

Progress has not been rapid in improving the texture and flavor of winter squash. Some varieties have a ropy texture, and their somewhat bland flavor has to be enhanced by baking with honey, brown sugar, or sausage. A few varieties are tender, but not mushy, and so sweet that only butter needs to be added for baking.

All squashes are native to the new world and, in North America, were grown among corn plants by Indian Tribes. Some seedsmen list the cushaw among winter squashes. This book lists it with pumpkins. Vegetable sphagetti is listed among the winter squash because it can be stored.

SITE

Winter squash needs a garden site with full sun, warm soil and good air circulation. Also, since most winter squash varieties have very long vines, you should choose your site carefully. Unlike cantaloupes and watermelons, winter squash fruits will remain attached to the vine when they are ripe. You can train the vines of small-fruited varieties up and over garages and sheds, and even up into trees if you are sufficiently athletic to retrieve the fruits.

The large-fruited varieties should always be run on the ground because the weight of the fruits can tear the vines loose from the support.

Another approach is to emulate the American Indian fashion of planting winter squash among rows of corn. The corn stalks will begin to die back in time to permit good late growth on winter squash, and an acceptable crop of fruits will be set. The corn stalks keep the rambling vines somewhat contained within the corn patch.

SOIL

Winter squash is only moderately tolerant of soil acidity, and will grow best at pH levels of 5.5 to 6.8. Winter squash needs soil which will retain a good reserve of moisture and nutrients throughout its prolonged growing period. Therefore, medium to heavy soils are better for winter squash, especially if your garden is so large that irrigation is impractical.

SOIL PREPARATION

The bush varieties of winter squash are usually planted on raised beds, but no such special preparation is needed for the rambling plants of vining varieties. Loosen the soil deeply, incorporating organic matter to a depth of 12 to 18 inches in order to provide a good, deep root zone for winter squash.

Large plots of winter squash are occasionally grown and, for these, green manure crops should be considered. The cover crop can be turned under in late spring and given about a month to decompose before planting squash seeds.

PLANTS PER PERSON/YIELD

Grow 2 to 3 plants for each person to get 12 to 15 fruits weighing from one-half pound to five or more pounds, depending on the variety. Vining types are the most productive, but take the most space.

SEED GERMINATION

Squash seed germinates best at soil temperatures of 70 to 95°F., but sometimes the emerging seedlings have difficulty in surviving in the hotter soil ranges. At 80°F., summer squash seeds will germinate in 3 to 5 days. A heavy crust of clay soil can retard germination.

DIRECT SEEDING

The usual way to plant squash seed in the garden is to sow them after all danger of frost is past and the soil temperature has warmed to at least 60°F. Mark off rows 24 inches apart for bush varieties and 72 inches apart for vining varieties. For bush varieties, place a group of 2 or 3 seeds every 18 inches apart, and for vining varieties every 24 to 30 inches apart. Scoop shallow depressions the depth and size of a saucer and place the seeds around the edge, covering them ½ to ¾ inch deep with sand, vermiculite or sifted compost.

In zones 1 and 2, where cool summers can cause cliffhanging situations if the first fall frost comes early, gardeners can gain 2 to 3 weeks on the season by planting winter squash seeds through a plastic mulch laid down about a month in advance to trap solar heat. Adhere to recommended planting dates because early planting of squash can be quite risky. The soil temperature under a clear plastic mulch can run significantly higher than that under black plastic mulch or on open soil.

If your squash crop is threatened by a late spring frost, be sure to cover seedling plants with hot caps, plastic canopies or cloche.

TRANSPLANTING AND THINNING

Some gardeners find transplanting helpful where growing seasons are short. Start seeds indoors in peat pots 3 to 4 weeks prior to the indicated direct seeding dates. Place two seeds per pot and cover ½ to ¾ inches deep. Grow young plants in full sun or under strong light from fluorescent lamps to prevent stretching. Harden the seedlings for several days before planting in the garden, and do not disrupt the roots when setting the plants into the soil. Thin groups or hills of squash to the two strongest seedlings per location.

SUPPORTING STRUCTURES

The small-fruited vining types of squash can be grown on sturdy fences or trellises four feet or more in height. Don't try trellising large, heavy-fruited varieties such as banana and hubbard types as their weight could pull down the structure or cause vines to break and fall.

WATERING

Fortunately, winter squash is rather drought resistant. Once the vines begin to ramble, watering by furrow irrigation becomes difficult if not impossible, and watering by sprinkling is not recommended because it dampens the foliage and increases the possibility of diseases. An ideal arrangement is either drip irrigation or an old soaker hose turned over so that the holes face down. The irrigation tube or soaker hose should be laid down the row while the plants are still small. If no rain falls for 10 to 14 days, let the watering system drip for 30-60 minutes daily. An adequate supply of moisture, once the plants have begun to bloom, is essential to a heavy set of squash fruits.

RECOMMENDED MULCH

Bush varieties are usually mulched with black plastic, but it is difficult to mulch the large areas which will be covered by the vines of running varieties. Some growers provide a "living mulch" by alternating cultivated rows of winter squash with wide bands of a green manure crop such as rye grass or one of the summer legumes. The green manure crop will keep down weeds and will not be hurt by the squash vines which will cover it by the end of the season.

You will see lots of exposed soil between the widely spaced young plants of winter squash and should keep these areas hoed clean of weeds until the plants grow together to cover them. Most gardeners don't bother with turning back the vines to hoe weeds and grass, but learn to accept them gracefully. Attempting to walk among the network of winter squash vines in order to pull weeds can crush and kill so many runners that any gain is cancelled.

CONTAINER GROWING

Only the bush varieties of winter squash are suitable for growing in containers. Even they are so large that only one plant can be grown per ten gallon container. Because of their high transpiration rate and high nutrient needs, plants require frequent watering and feeding. A compact variety such as Gold Nugget is ideal for container growing; you might try two plants of Gold Nugget per container if the tub is rather shallow and broad, giving sufficient room for the two plants.

ENVIRONMENTAL PROBLEMS

Because of their vigor, winter squash are not subject to as many problems as summer varieties. Failure to set a heavy crop of fruit is a common problem, and this is usually due to long periods of cool, damp weather in late summer or to the absence of a sufficient number of bees to provide good pollination. If you live in the North or in cool coastal areas it is suggested that you use the earliest varieties of winter squash. These have the ability to set fruit at somewhat lower temperatures than the later varieties. Good fruit set can often be promoted by starting plants early indoors in northern areas and setting them out under protection in early summer. Plastic canopies will bring along the young plants very quickly and can result in a first fruit setting two to three weeks earlier than they would normally.

HARVESTING

A key to successful storage is to harvest winter squash fruits at the correct time after they mature. Skin color and toughness are prime indicators of maturity. Butternut types change from green to tan, while the skin of most green varieties develop bronze, orange or brown areas. Spaghetti squash will develop a deep rich yellow skin color. A simple and universal test for all winter squash is to test the skin with the thumbnail — the skin should be hard enough to resist scratching or puncturing easily.

The stalk which attaches the squash to the vine is another guide to harvest readiness. It should be fairly hard and woody. On some varieties it will be shriveled, with noticeable ridges. The correct way to harvest is to cut the fruits from the vine. Cut the stalk with shears, leaving a piece about one inch long attached to the fruit.

To a limited extent, you can also use the calendar as a guide to harvest time. Winter squash require eighty to one hundred growing days to mature. However, "days to maturity" can vary considerably, depending on whether your area has a warm or cool summer.

Usually, the majority of squash fruits are ready for harvest at the same time. Their sugar content increases after some exposure to chilly weather. Often the vines will be killed by a light frost, but don't wait until a hard freeze hits. The frozen fruits will not store well. Try to harvest on a sunny day, after a spell of dry weather. Handle the fruits carefully; do not wash them prior to storage, but brush off soil gently.

STORAGE

Cure well matured winter squash at temperatures of 80-85°F. and a relative humidity of 80-85% for 10 days. Curing hardens the skin and heals surface cuts. After curing, store in a cool, well-ventilated area at 50-60°F. The humidity in the storage area should be high but not so much as to cause condensation. In an excessively dry storage area, winter squash will lose weight and food value rapidly.

The storage life of winter squash under good conditions is several months, with the exception of acorn and banana squash which will keep for only about two months. Watch for a color change on green varieties of acorn squash; when they turn orange they will have lost moisture and will have become stringy.

STORING LEFTOVER SEEDS

Squash seeds are among the longest lived of any vegetable. When placed in a sealed container with a desiccant and stored in a refrigerator, seeds should germinate at least 50% after five years.

GLOSSARY OF TECHNICAL TERMS

ACIDITY OF SOIL -- measures the concentration of hydrogen ions in the soil expressed as pH. Garden soils that are described as "acid" are usually in the pH 5.0 to 6.0 range and are most often found in rainy areas where the soil is subject to leaching and the neutralizing agents have washed away. Some acid soils are derived from acid-forming baserock or aquatic plants. Acid soils are usually deficient in calcium and magnesium and often in phosphate.

ALKALINITY OF SOIL -- measures the concentration of hydroxl ions in the soil expressed as pH. "Alkaline" soils are usually in the pH 7.5 to 8.5 range. Such soils are common in the arid West and, elsewhere, in soils derived from limestone baserock. Alkaline soils are often deficient in available phosphate and available iron.

ANNUAL -- a species that completes its life cycle, from planting through flowering and setting seeds, in one growing season. Bush beans are a good example.

ARTIFICIAL SOILS -- are designed for container growing of vegetable, flower and woody plants. The first was the "University of California Mix" developed for nurserymen. It was composed of a mixture of sawdust or shavings, and sterilized sand. Limestone and starter fertilizer were included. Some sterilized soil was occasionally included. Later, the "Cornell or Peatlite" mix was developed. It contained peatmoss and vermiculite or Perlite, starter fertilizer and lime, except when used with acid-loving plants where gypsum was substituted.

Even later, mixes containing pulverized pine bark in lieu of the more expensive peatmoss were developed in the South.

Artificial soils take water easily, drain rapidly or fairly rapidly and do not shrink away from the sides of the container when dry. They are less likely than garden soils to harbor diseases.

BANKING -- blanching the lower stalks of leafy vegetables by enclosing them within boards or by banking up the plants with loose, dry soil, straw or hay. See "Celery".

BEDS OR SEEDBEDS -- are the garden areas between footpaths. Beds may be raised or "on the flat". Beds are usually wide enough to accommodate 1 or 2 rows of plants but not more than 4 ft., the practical limit for easy reach for weeding or harvest.

BIENNIAL -- a species that completes its life cycle in 12 to 18 months. A good example would be parsley which, if spring planted and protected to winter over, would flower the following spring and set seeds in the summer. Plants would die after maturing seeds.

BITTERNESS -- an inherited characteristic made worse by drought stresses. Encountered in cucumbers, endive, summer lettuce and, occasionally, squash.

BLOCK PLANTING -- used for sweet corn to get better cross pollination than is possible from single or double rows.

BOLTING -- occurs when a plant begins to go to seed. It shoots up flowering stalks which form seeds. In some plants bolting is triggered by cold periods while in others it results from changes in night length. Bolting-resistant varieties are available in many species.

BOTTOM HEAT -- the most effective way to maintain the optimum temperature within growing media to produce plants. Precisely controlled bottom heat can speed seed germination and give improved seeding performance. Heating cables and mats are available.

BROADCASTING SEEDS -- means scattering seeds across the full width of the prepared seedbed. Usually done with seeds of salad greens and potherbs such as collards, mustard greens and turnips. Broadcasting is most successful on improved soils with few weeds and good water penetration.

BUTTONING -- the formation of tiny heads of cole crops on small plants, caused by the late transplanting of overly-mature seedlings.

CAGES -- structures of wire or plastic fencing or reinforcing wire, used to enclose and support plants of tomatoes, vine crops and other climbing plants.

CAPILLARY ACTION -- water can move up through the soil by a physical force known as capillarity. Pores between soil particles interconnect to form channels through which water is pulled. Deep and thorough soil preparation can reduce interference with capillary movement by breaking up layers of soil amendments or deposits called "hardpans."

CHECKING OR GROWTH CHECKING -- occurs when a rapidly growing plant is stressed by drought, cold, excessive soil moisture or by the shock of tranplanting. Growth is slowed and development stalls, sometimes with serious consequences.

CLEAN CULTIVATION -- keeps the soil free of weeds and the surface soil loose by hoeing or shallow tilling. No mulches are used.

CLOCHE -- a device long used in Europe to protect early planted seedlings and late vegetables from frost and cold winds. The traditional cloche is an A-frame structure of glass panes clamped within heavy wire. Cloches can be overlapped to cover rows or used singly by closing the ends.

COMPOST -- books have been written about the benefits of compost and composting. Compost is decomposed organic matter somewhere between raw vegetable matter and fine particles of humus. It is produced by soil organisms working on vegetable matter in a warm, well-aerated, moist environment, neither excessively acid nor alkaline. "Mature" compost is nearly odorless and crumbles easily. An active compost heap can generate enough heat to kill many, but not all, weed seeds. Compost is considered a soil amendment, not a fertilizer, and is valued principally for loosening soil while providing improved drainage, moisture and nutrient retention and biological activity.

CROP ROTATION -- can refer to growing a given species in a different part of the garden each season, or to the procedure of interspersing green manure crops between vegetable or grain crops to maintain soil structure and fertility.

CROSS-POLLINATED -- refers to the species or named varieties which require, for seed set, that pollen be transferred by wind or insect activity. Geographical isolation of seed fields is especially important to avoid crossing with closely related members of the same species.

CRUST -- a dense layer of soil that can form over seedbeds and inhibit germination and growth. Crusted soils are difficult to water, weed and cultivate. Crusts are prevalent on clay soils, especially those low in organic matter.

CURD -- the clustered flower buds that make up the heads of cauliflower.

DESICCANT -- a material used to absorb moisture from seeds, enabling them to remain viable longer. Typical desiccants are silica gel and calcium chloride. Powdered milk can be used in a pinch.

DIRECT SEEDING -- planting seeds in the garden as opposed to starting seeds indoors or in an outdoor nursery bed for transplanting. Direct seeded plants usually are allowed to remain where they take root, but surplus seedlings of most species can be transplanted.

DOUBLE-DIGGING -- the process of digging and laying aside successive layers of garden soil to a depth of 18 to 24" and, preferably, incorporating organic amendments before returning the soil to the trench in the reverse order of its removal. Double-digging is not recommended where heavy clay subsoil underlies good loam, nor is it recommended for soil to be used for growing sweet potatoes.

DRILLING -- planting seeds close together in a row. Fertilizers can also be drilled in furrows to the side of seed rows. Drilling improves the utilization of fertilizer ingredients.

DRIP IRRIGATION -- trickle or drip irrigation was perfected in Israel to make best use of scarce water in desert areas while increasing production. Numerous variations have been developed, including porous tubes that "leak", tubes with small, regularly spaced holes, and tubes with special emitters. A properly installed drip system can result in significant increases in production, in labor and water savings, and in saving crops during periods of dry weather.

FRAMES -- enclosures of lumber in 2x4 to 2x12 sizes, or logs, cinderblocks, short posts or other devices to keep soil elevated in beds. Frames are usually filled with soil modified with amendments and are most useful on poorly drained soil or for root crops.

FURROWS -- V-shaped ditches cut with a hoe or plow. Deep furrows are often used as footpaths between beds. Shallow furrows are made to plant rows of seeds.

FURROW IRRIGATION -- much used in level-land agriculture where water can be directed down furrows between rows of vegetables. A slight slope keeps the water flowing slowly, with a minimum of erosion. Popular for deep watering in western gardens. Furrow irrigation does not encourage the spread of foliage diseases and results in less loss to evaporation than sprinkling.

GENE POOL -- the collective inheritance of crops, passed down through many generations. Some species are genetically stable and breed true year to year. Others require many years of selection and periodic renewal to produce uniform plant populations.

GERMINATION -- occurs when the root or shoot, or both, have pierced the seed coat and have entered the soil. Germination can occur days before emergence, at which stage you can see the seedling emerging from the soil. The "percentage of germination" refers to the laboratory performance of seeds and is the percentage of normal seedlings that sprout from a given lot of seeds. "Federal Minimum Standard" is a compromise figure reached between seedsmen and seed enforcement officials. It represents the percentage of germination you can expect, under good growing conditions, from an average lot of seeds that has not been subjected to excessive heat, humidity or mechanical injury or to long storage. Any lot of vegetable seeds that does not meet FMS must be prominently labeled "Below Standard".

GREEN MANURE CROPS -- are grasses, legumes or heavy-yielding leafy crops that are grown to be turned under for organic matter. Experienced gardeners plant seeds of green manure crops among vegetables late each season.

GYNOECIOUS -- an inherited trait for nearly or all-female blossoms. See "Cucumber" for a complete discussion.

GYPSUM -- powdered calcium sulfate, used principally to granulate clay soils to make them workable. Also used as calcium source when it is undesirable to raise the pH.

HARDENING OFF -- or acclimatizing a plant reduces the shock of moving it from a relatively warm, protected environment to a site where it may have to cope with drying winds, often chilly; cold, moist soil and strong sunlight. Hardening consists of gradually exposing plants to the new stresses they must withstand.

HARDINESS -- a much abused term, the source of considerable confusion. If a plant is described as "hardy", try to ascertain if the writer means winter hardy or if he or she is using hardy to infer durability under stresses such as heat and drought.

HARDPANS -- lenses or layers of concrete-like particles and salts lying at various depths in the soil. These usually form where dissolved salts or minute particles, when flowing down through the soil, encounter obstructions. Plowsoles are a special kind of hardpan formed by mechanical compaction such as occurs in plowing. Poorly drained or excessively dry spots in a garden may be caused by hardpans at a depth of 10 to 30".

HEAVY SOIL -- a term used to describe clay or clay loams that are dense and heavy when wet.

HERBS -- special plants grown for seasoning, bathes, medicinal use, scenting, and repelling insects.

HILLING -- originally a farming term, used to describe planting seeds such as corn in regularly spaced groups so that fields could be cultivated down the rows and across them as well. This threw soil up around the bases of plants in "hills". Now the term is also used to describe planting on low mounds or "hills" for improved drainage and warmer soil.

HYBRID VARIETIES -- the result of crossing plants, usually within the same species, to achieve greater vigor, uniformity and production than would come from any of the parents. Hybrid seeds are expensive to produce but are usually worth the additional cost. If you save and plant seeds from a hybrid you grow in your garden, the offspring will be variable because of the mixed parentage. Modern plant breeding techniques are now producing crosses between separate species and, in rare cases, between separate genera.

INOCULATION -- treating seeds or the soil in a planting furrow with spores of nitrogen-fixing bacteria. Inoculation helps peas and beans grow better on new soils.

INTENSIVE GARDENING -- growing more in less space by maintaining optimum levels of soil moisture and fertility, sunlight and biological activity. Special practices include close spacing, prompt and frequent harvesting, succession planting and interplanting, mulching and careful choice of varieties.

INTERPLANTING -- usually means planting fast growing species in the bare areas left for the expansion of slow growing plants with large plants. The fast crops can be harvested before they crowd the later vegetables. Caution: interplanting among vine crops is tricky as the vines tend to overrun the fast crops before they are ready.

LEACHING -- occurs when water dissolves or hydrolyzes chemical compounds and carries them down through the soil and out of reach of plant roots. Leaching may strip soil particles of nutrients while depleting fertilizers. Leaching is more rapid and severe on sandy soils in rainy areas and on artificial soils in containers.

LIGHT SOIL -- describes sandy soil or sandy loam soils that are comparatively lightweight when dry.

LIMESTONE, GROUND -- may be either calcium carbonate or a mixture of calcium and magnesium carbonates called "dolomite". Farmers can buy various grinds of limestone but usually only medium sized mill run limestone is sold to home gardeners. Limestone releases relatively more calcium and magnesium the second year after application. Limestone is not used on arid western soils where, because of limited leaching, supplies of calcium and magnesium are usually adequate.

LOAM -- a type of soil containing a mixture of sizes of soil particles from sand or silt to fine clay. Loams range from open, sandy loam to rather tight clay loam.

MIDGET VEGETABLE VARIETIES -- these have genetically controlled plants of smaller size than standard varieties within the same species. The fruits of midget varieties may be either miniature or of standard size; read descriptions carefully.

MILDEW -- a plant disease characterized by dusty or downy patches of fungal growth on leaves, stems and fruit. The universal form is powdery mildew which is often seen on vine crops late in the season. Sprinkler irrigation encourages the spread of mildew.

MONOGERM -- some kinds such as beets and swiss chard naturally produce seeds in clusters called " seed balls", which result in plants coming up in clumps. "Monogerm" varieties produce only one seed per seed ball and can be planted with greater precision.

MOSAIC -- Races of virus diseases of plants, affecting many species, are collectively called "mosaic". A few viruses are seed-borne but most are transmitted by the feeding of insects. The mottling of foliage, distortion of stems and fruits and stunting are usually less severe on fertile, well-watered soils. Chemical control of plant viruses is expensive and difficult. Once severe mosaic disease is confirmed, gardeners usually pull out affected plants, not touching healthy plants in the process.

MUDDING -- the soaking of bare-root seedlings in muddy water to revive them and to coat root hairs with clay mud to reduce transplanting shock.

MULCHES -- mulching controls soil temperature, conserves moisture and keeps down weeds. Mulches may be of many kinds of vegetable litter, paper or cardboard, or plastic sheeting.

NIGHTLENGTH RESPONSE -- the regulation of growth habit of various species by hormone levels created by changes in the number of hours of darkness. Typically, most species tend to flower and set seeds when nights grow shorter. Nightlength response can often be traced to the geographical latitude where the ancestral plants originated.

NITROGEN, AMMONIACAL FORM -- many fertilizers, especially organic, contain all or mostly ammoniacal forms of nitrogen that require bacterial conversion to nitrates for availability to plants. Molecules of ammoniacal nitrogen are attracted to soil particles and leach rather slowly. However, they are toxic to plant roots when present in excess, such as can happen in cold soil with a low level of bacterial activity.

NITROGEN-FIXING -- certain plants, mostly in the legume family like beans and peas, can get by with less supplementary nitrogen than other plant species. Colonies of beneficial bacteria cluster around the roots and cause nodules to form. These nodules are benign and do not interfere with nutrient or water uptake. The bacteria can extract nitrogen from the air and convert it into forms that can be absorbed by plant roots.

NITROGEN, NITRATE FORM -- plant roots, except those of a few specialized species, can absorb only nitrate nitrogen. Some fertilizers have to be worked on by beneficial soil bacteria to be converted to the nitrate form. Nitrates are repelled by soil particles. Hence, they slip through the soil and quickly leach away unless captured by feeder roots.

NUTRIENTS, PLANT -- are mineral elements necessary for growth and development. They may come from the soil itself, from the decomposition of amendments such as manure or other organic matter, or from natural or manufactured fertilizers. Most fertilizers contain only the three major elements: nitrogen, phosphorus and potassium. Take note that certain soils need supplementary applications of the secondary nutrients such as calcium, magnesium and sulphur, and micronutrients such as iron, copper and boron.

OPEN POLLINATED -- plants that are allowed to produce seeds with no manipulation other than isolating seed fields which might cross with other varieties of the same species. Open pollinated seed production include species with self-pollinating flowers as well as those requiring agents such as wind or insects for pollen transfer.

ORGANIC MATTER AND HUMUS -- in soil, organic matter decomposes into fine dark particles called "humus". Organic matter may come from any kind of vegetation, animal wastes or by-products. Any solid matter except minerals can come under this broad classification.

RIDGED-UP ROWS -- high, narrow beds between closely spaced footpaths are called ridged rows. These are often seen in the South where spring crops have been planted on heavy, slow draining soils. Ridged-up rows are not recommended for sandy soils or for summer crops on heavier soils because the soil tends to run too hot and dry.

ROASTING EARS -- immature ears of field or grain corn, harvested in the milk stage for boiling or roasting. Long an alternative to sweet corn in the South.

ROWS -- lines of plants , straight or contoured around the slopes. Single-row plantings, where rows are separated by rather wide footpaths, do not make efficient use of garden space. Row plantings make it easier to distinguish vegetable seedlings from weeds than would be the case with broadcast plantings.

PARCHMENT -- the sheet-like fiber found in pods of unimproved bean and pea varieties.

PARTHENOCARPIC -- fruit formation without pollination. This is an inherited trait. See "Cucumber" for more information.

PEATMOSS -- can include products derived from decomposed sedges and woody plants as well as from true sphagnum peat. The products can vary widely in physical and chemical properties between brands. The best peatmoss for general garden use is coarse sphagnum peat; the worst is the black muck that drains slowly and often contains undesirable minerals and weed seeds.

PERLITE -- the trade name for an expanded silica ore much like the pumice produced by volcanoes. It is very light in weight and sterile, but holds little water. Perlite is often used in potting soils to improve drainage and reduce weight.

PERENNIAL -- not many vegetables will regrow from the same root system for several years, the nature of perennials. Asparagus and rhubarb are examples of winter hardy perennials. A tender perennial which lives over in zones 6 & 7 would be chayote, little known further north.

PLANNING VEGETABLE GARDENS -- this book does not offer advice on laying out plans for vegetable gardens. However, for planning purposes, you will find that the information on amounts of vegetables to plant per person, average sizes of plants, and planting dates given in this book will give you a good start on planning. Your local County Agricultural Extension Service offers help in planning.

POTHERB -- a leafy vegetable grown principally for cooking, not to be confused with culinary herbs which are grown principally for use as seasonings and garnishes.

PRE-SOAKING AND PRE-SPROUTING SEEDS -- see the various kinds of vegetables. These seed treatments have proved helpful in improving the speed and percentage of germination of certain kinds of vegetables.

SALTS IN SOILS -- salts are commonly based on calcium, magnesium, sodium, iron and aluminum and are deposited where water carrying dissolved salts evaporates. Salts may come from the soil, from water, or from fertilizer. They are not usually troublesome except on arid western soils and on soils that are allowed to go dry after excessive applications of fertilizer. High levels of salts in the soil can kill plant roots.

SAVOY -- a term used to describe the crumpled, textured leaves of certain varieties of leafy vegetables such as spinach and cabbage.

SEEDLING VIGOR -- can vary from lot to lot of the same variety of seeds, and is influenced by a number of growing, harvest and storage conditions. Seedling vigor is distinct from germination but can affect it. In some species, the size of the seeds is important to vigor. Vigor is most important in early planted seeds when the soil is cold and wet.

SEED PACKETS -- a seed trade term used to describe sealed envelopes containing enough seeds to grow vegetables for an average sized family. "Trade Packets" are larger. "Bulk Seeds" are sold by weight.

SEED STORAGE -- seeds will keep longer if placed in a sealed container with a desiccant, and kept in a refrigerator at about 40°F. Storage of seeds under high temperatures and high humidity can kill them in as little as 30 days.

SEED TREATMENT -- seeds can be treated with fungicides and/or insecticides to improve germination. Fungicidal treatments are often used with early planting in cold soils. Some catalogs offer treated seeds but most don't because of the objection of organic gardeners to chemical treatments.

SIDE SHOOTS OR LATERALS -- the smaller secondary heads or buds of cabbage or broccoli that form on certain varieties after the main crop has been harvested.

SOIL AMENDMENTS -- are materials worked into the soil to alter its physical texture and structure. Amendments may contibute small amounts of nutrients to the soil, but their main purpose is to improve drainage, water and nutrient retention, biological activity and ease of cultivation. Soil amendments include such materials as sand, compost, manure, peatmoss, sawdust or shavings, etc. Limestone, gypsum and agricultural sulphur are considered soil amendments. Amendments are not the same as fertilizers, which are concentrated sources of plant nutrients.

SOIL COMPACTION -- takes place from the weight of foot traffic or machines. Compaction sqeezes the air out of soil and makes the soil dense, a condition particularly resented by root crops such as carrots and onions.

SPECIES -- a term not often used in the seed trade because it refers to distinct plant populations in the wild, or cultivated with little attempt at selection. Most vegetables now only remotely resemble their wild parent species.

SUCCESSION PLANTING -- growing two or more crops per season in the same space. Succession planting and interplanting meet when replacement crops are planted alongside maturing vegetables in order to save time.

SPACE-EFFICIENT -- a term used to describe vegetables that produce high yields of valuable food per unit of time and per unit of garden space. Tomatoes and peppers are good examples of space efficient vegetables; pumpkins and sweet corn would be inefficient.

TAPROOT -- some vegetables have a strong central root that penetrates deep into the soil. These species are usually difficult to transplant, as compared to plants with a spreading mat of lateral roots.

TEEPEES -- a general term covering pyramidal structures such as tripods or quadripods of lath, lumber, poles, pipe or bamboo — spread at the base and lashed together at the top to provide support for climbing vines.

THINNING -- pulling out or snipping off excess plants to provide the proper spacing between vegetable plants for optimum growth and production.

TRANSPLANTING -- moving a plant from one site to another and setting it in the soil where roots can grow and spread. Transplanting may be done indoors, when plants are removed from a seedling flat and set into individual pots; from indoors to outdoors, when setting started plants out in the garden; or from outdoors to outdoors when moving surplus direct seeded plants.

TUBERS -- not roots, tubers are swollen underground stems. Examples are Irish potatoes and dahlias.

VARIETIES -- an old term, still favored by the seed trade to describe distinct, named selections from a given species or from a hybrid ancestry. Plants grown from seeds of a given variety will be quite similar but not virtually identical as are those produced by cloning a single plant. Modern nomenclature leans toward the term "cultivar", meaning cultivated variety. Certain old varieties include several "strains" resulting from the selections made by different plant breeders. Strains may differ slightly, but when they digress too greatly from the accepted standard, they should be renamed.

VERMICULITE -- a lightweight, sterile, water absorbent soil amendment manufactured by expanding mica ore under heat and pressure. Much used in artificial soil mixtures and for sprouting seeds in seed flats. Contains some potash.

VERNALIZATION -- exposure of plants to a certain number of days of relatively cold temperatures to improve production. Vernalization should be avoided in plants which are prone to bolting, and for warmth loving species.

VINE CROPS -- include cucumbers, melons, pumpkins, watermelons, squash and minor crops such as chayote and citron.

WIDE BAND OR BROAD BAND PLANTING -- is direct seeding in bands of up to 24" width to gain increased production of leafy vegetables and fruiting types with compact plants. Clean, weed-free soil is an asset when planting closely in bands.